WORLDS OF WELFARE

Throughout the world welfare systems have seen a period of unprecedented change. Understanding these changes is difficult, not least because they vary so much from place to place.

Worlds of Welfare provides a clear and concise guide to these changes. Part I examines the various types of welfare state around the world and describes the numerous reforms – such as privatisation and commercialisation – that have been introduced in recent years. Part II tests the many theoretical perspectives for understanding such social change. Steven Pinch concludes his comprehensive study with a look into the future of the welfare state in multicultural societies.

Thanks to the clarity of Pinch's writing and an extensive glossary of key terms, *Worlds of Welfare* makes an impressive case for the importance of geographical perspectives on welfare reform. It will be of interest to all concerned about the future of welfare services.

Steven Pinch is Senior Lecturer in Geography at the University of Southampton.

WORLDS OF WELFARE

Understanding the Changing Geographies of Social Welfare Provision

STEVEN PINCH

Routledge
Taylor & Francis Group

LONDON AND NEW YORK

First published 1997
by Routledge
4 Park Square, Milton Park, Abingdon, Oxon OX14 4RN
605 Third Avenue, New York, NY 10017

*Routledge is an imprint of the Taylor & Francis Group,
an informa business*

British Library Cataloguing in Publication Data
A catalogue record for this book is available from the British Library.

Library of Congress Cataloguing in Publication Data
Pinch, Steven.
Worlds of Welfare : Understanding the changing geographies of social
welfare provision / Steven Pinch.
p. cm.
Includes bibliographical references and index.
1. Public welfare–Cross-cultural studies. 2. Welfare state–Cross-cultural
studies. 3. Social policy–Cross-cultural studies.
I. Title.
HV37.P558   1996
361–dc20   96–7594

ISBN 13: 978-0-415-11189-8 (pbk)
ISBN 13: 978-0-415-11188-1 (hbk)

CONTENTS

LIST OF FIGURES

LIST OF TABLES

PREFACE

This book emerged from an invitation to write a second edition of my earlier book, *Cities and Services*. Upon contemplating this daunting prospect, it soon became clear to me that so many developments had taken place in the study of geographical aspects of welfare services in the interim since *Cities and Services* was conceived, that a completely new volume was needed. These changes involve not only the character of welfare systems but also the ways in which these systems are studied. The two elements are of course interrelated. *Cities and Services* was written between 1982 and 1984 but much of the material was derived from data and frameworks stretching back to the 1960s and 1970s. During this post-war era many features of the welfare state were relatively stable and there was much confidence that scientific analysis, planning and social engineering could be used to overcome many of the problems generated by social democratic capitalist societies. In the early 1990s there is no such confidence. In the wake of the economic problems experienced in the mid-1970s, welfare systems in the advanced industrialised nations have been subject to enormous changes in recent years. Furthermore, there is considerable uncertainty over the advantages and disadvantages of these changes. In a parallel fashion, the confidence that surrounded both positivist and Marxist theories in earlier years has evaporated.

This book is therefore concerned both with the changing geography of the welfare state and with the changing nature of geographical approaches to these shifts and so is complementary to *Cities and Services* rather than a direct replacement. There is a two-part structure; the first part examines some of the numerous changes that have taken place in the structure of the welfare state and their geographical manifestations, whilst the second part looks at some of the explanations that have been advanced to account for these changes. Many of the changes in the welfare state have been distressing for those involved, whether as producers or consumers. Yet the story is not entirely one of gloom, for there have also been many progressive developments. Similarly, on the academic side, whilst there are negative developments, substantial progress has also been achieved in many quarters. Indeed, the last ten years have been an extraordinarily dynamic time in geography and I have tried to convey some of this momentum. However, no book can do everything and I have deliberately avoided certain large areas in which there is already a substantial literature. Thus, I have avoided the huge amount of material on the history of welfare states and some of the excellent work in specialist fields such as work on the British NHS (see Mohan, 1995) and social justice (Smith, 1994). I have also deliberately not attempted to evaluate in great detail the merits of the various criticisms of the welfare state or the success and failure of the various reforms. However, it should be clear from the exposition that many of the existing critiques of the welfare state are deeply flawed. Furthermore, so far, there is little evidence to support many of the recent changes in welfare systems.

A final guiding principle is that, as in *Cities and Services*, I have tried to convey the ideas in an accessible manner, using technical jargon only where it is necessary to understand the basic ideas. Academic writing inevitably spawns a vast complex vocabulary which can seem intimidating at first glance. However, some of the ideas are less vicious upon closer acquaintance. I have therefore included a glossary (see page 138) to act as a guide through the terminology. This glossary can be used in two ways: first, as a way of tracking down a particular concept; and second, as a memory aid after reading a particular section of the book. There is also a guide to further reading after every chapter.

ACKNOWLEDGEMENTS

One of the great pleasures of completing a book is the opportunity it provides to thank formally all those people who have helped one over the years during which it was in preparation. First and foremost, I must thank Tristan Palmer of Routledge for his advice and encouragement. Without this I would have surely abandoned the project and published more of those international refereed journal articles that have become such vital elements in the higher education system in recent years. Thanks must also go to Paul Knox and the anonymous reviewer for their helpful comments on the first draft. I must also thank those who took the trouble to write and give me feedback on *Cities and Services*, since some of these ideas I have incorporated into *Worlds of Welfare*. Thanks also to Barbara Crow for her excellent bibliography of feminist literature.

On the academic front, however, giving thanks presents a problem for there are so many individuals to whom one is indebted. Whatever it may appear to the contrary, academia is inherently a cooperative endeavour; indeed, I often think that British geography is like some huge 'industrial district' or 'local world of production' with various individuals contributing towards, and benefiting from, the creation of a 'knowledge trajectory'. Certainly, there has been an enormous amount of new and stimulating work in geography in recent years. For example, in the sphere of public services I have learnt much from scholars including Alan Cochrane, Mark Exworthy, Clive Forster, Mark Goodwin, Susan Halford, John Humphries, Rob Imrie, John Mohan, Graham Moon, Joe Painter, Alan Patterson, Jamie Peck, Alan Phipps, Philip Pinch, John Stubbs, Molly Warrington and many, many others whose excellent works are listed in the bibliography.

Nearer to home, although I have remained in the same department in the last decade, I could not have experienced greater change if I had gone elsewhere. A hurricane of reform has swept through Southampton Geography Department and has brought in (and at times swept away) a set of hugely talented young geographers from whom I have learnt an enormous amount: Alison Blunt, Nick Henry, David Martin, Andy Merrifield, David Pinder and Jane Wills. My long-standing academic colleague and friend Colin Mason has also been a great strength, not least through his example of industry, decency and integrity. I have also learnt much from my post-graduate students: Nick Goodwin on contracting-out and NHS reforms; Howard Hurd on corporate philanthropy; Jon Chipp on post-modernism, post-structuralism and post-everything else; William May on flexibility (and for helping me to obtain references); David Wright on sexuality and welfare; and Kristie Legg on issues of exclusion, knowledge and power. I must also thank Tim Apsden and his cartographic staff at Southampton for their highly professional service.

Finally, getting to home, there have been even more changes. *Cities and Services* was disrupted by pleas to fly kites, build with Lego and cook dinner. Alas, the first parts of this book were written whilst my offspring were in late adolescence and there was little to break the silence apart from the sound of Oasis through my study walls (plus the demands for dinner). Nevertheless, through their interminable questioning of my intellectual legitimacy (not difficult) and perpetual banter, I have learnt much from my sons David and Richard. Finally, thanks to Lyn for being just Lyn and for goading me to 'finish the bloody thing'.

Part I

CHARTING THE DIMENSIONS OF CHANGE IN WELFARE STATES

1

INTRODUCTION

Descending through the skies towards an airport, the first sight of a new country through an aircraft window often brings a feeling of familiarity. Viewed from an aerial perspective, most cities in advanced industrial societies seem broadly similar: high-rise office blocks huddle in downtown areas symbolising the corporate power (or indebtedness) of their occupants; factories and warehouses cluster in zones; highways snake towards the urban periphery; and sprawling low-density suburbs combine with areas of higher-density flats and apartments. Even after landing and observing the local inhabitants, the visitor is often struck by the things that unite peoples – at least in the more affluent parts of the world. Not the least of these things held in common are consumer goods such as Japanese cars, US computers, Taiwanese electronic goods, French wines and Italian shoes. Furthermore, if able to communicate with some of the inhabitants of the city, the visitor is often struck by the ways in which people share common aspirations for more basic things such as adequate housing, education and health care. It is therefore hardly surprising that such basic aspirations were enshrined in the Universal Declaration of Human Rights following the creation of the United Nations in 1948 (albeit in sexist terminology):

Everyone has the right to a standard of living adequate for the health and well-being of himself and his family, including food, clothing, housing and medical care and the necessary social services and the right to security in the event of unemployment, sickness, disability, widowhood, old age or other lack of livelihood in circumstances beyond his control.

(United Nations, 1948, cited in Pierson, 1991, p. 127)

These basic needs are often summarised by the term 'welfare'. However, it does not take the visitor to a new country long to appreciate that welfare needs can be met in many different ways. Thus, some nations depend heavily upon care by families and friends; others rely upon voluntary and charitable bodies; others depend heavily upon private forms of care by businesses; while in some countries the state plays a crucial role. Throughout the world the extent to which people rely upon the state to meet their basic needs shows especially wide variations. For example, a middle-income person in Britain is likely to rely upon the state funded and state provided National Health Service (NHS), whereas a person of similar status in the US is more likely to possess a private health insurance scheme and use privately owned health care centres. Furthermore, it does not take the inquisitive visitor to a new country long to appreciate that there are considerable variations *within* nations in the ways in which welfare needs are met. For example, despite the widespread sale of public housing in Britain in the 1980s, public sector housing estates are still visually distinct from owner-occupied areas of British cities. In US cities, the poor of the inner areas are often forced to rely upon underfunded public hospitals which contrast with the excellent facilities to be found in suburban areas.

Despite this diversity of ways in which welfare needs can be met, a broadly similar set of arrangements have been established in most of the advanced industrialised countries which ensures that the state plays an important role in meeting welfare needs. These arrangements are generally known as the *welfare state*. As will be seen later in this book, the character and functions of the welfare state have been the subject of immense controversy. What is clear, however, is that following a period of about thirty years of relative stability after the Second World War, the welfare state is experiencing a period of rapid change.

This book is concerned with understanding the changing nature of these welfare states. Because the nature of welfare arrangements varies so much, both between and within countries, it is argued that a geographical perspective is essential to understand these changes. The analysis concentrates upon the nations of the English-speaking world – Britain, the United States, Canada, Australia and New Zealand. Despite substantial variations in the character of their welfare systems, these 'Anglo Saxon' economies share a number of basic similarities in the ways in which they have restructured their welfare systems in recent years. However, to avoid an Anglocentric bias, these arrangements in the English-speaking world will, at times, be contrasted with the welfare arrangements in other types of industrialised nation and, in particular, with arrangements in what are sometimes termed the 'integrated' economies of continental Europe and Japan.

This first chapter outlines some of the basic concepts that are important in understanding the changing geography of the welfare state. First, it considers what is meant by the concept of 'welfare' and outlines the diverse ways in which these welfare needs can be met. This diversity is illustrated by a description of welfare needs in the three main blocks that constitute the advanced industrialised nations: Europe, the United States and Japan. This is followed by a discussion of why geography is important in understanding the changing nature of welfare. The chapter concludes with an outline of some of the forces that have led to the restructuring of welfare states.

WHAT IS MEANT BY WELFARE?

In Britain and the United States the term 'welfare' is often synonymous with the provision of income or services by the state. The term is often used in a pejorative sense to suggest that individuals who are reliant upon the state are in some ways inadequate since they are lacking in self-reliance. However, the term is also used in academic circles in a more general sense to refer to basic human needs.

There are many ways of classifying these human needs. One of the most simple but influential ways of classifying these needs is according to Maslow's (1954) pyramid-like hierarchy. At the bottom of the pyramid are the basic physiological needs such as oxygen, fluids, food, waste disposal and rest. Once survival is assured by meeting these basic requirements, other needs come into play, such as the need for safety, comfort, love, belonging, respect and self-esteem At the top of the pyramid is self-actualisation – the ability to create and understand music, art and beauty. The crucial point about this hierarchy is that the needs at the top of the pyramid can be satisfied only once the more basic needs at the bottom of the hierarchy have been met. Although all humans require food, water, warmth and sexual activity for survival, there are a vast range of ways in which these basic needs can be met. Thankfully, most of the peoples in the advanced industrialised world live at levels above those of mere subsistence, and so the concept of need is inevitably relative in character, referring to the adequacy of existing conditions in relation to some socially acceptable norm. However, the ways in which people define needs varies, not only between nations but also between different groups within nations.

Fortunately, there is once again a simple but influential typology that draws attention to this variation in the ways in which differing groups define needs. Bradshaw (1972) makes a distinction between four types of need depending upon whose definitions are used. First, there are *normative* needs: those defined by administrators, managers and professionals such as those that occupy the bureaucracies of the welfare state. Second, there are the *felt* needs of the public which may or may not correspond to a third category of *expressed* needs, as revealed when people apply, complain, petition or demonstrate for a service. The final category is *comparative* needs, defined by measuring the characteristics of those in receipt of a service. The crucial point to be derived from this conceptualisation is that, though different, these concepts of need are all highly interrelated. Often, the greater the amount of a service that is provided (as defined by the normative needs of professionals) the more demand there is for the service (expressed needs); conversely, restricting publicity and supply can dampen down demand for the service. Generally speaking, the characteristics of those in receipt of a service (comparative needs) will reflect to varying degrees the values of administrators (normative needs) and the public (felt needs). Conflicts between these two groups have been a major source of controversy within the welfare state.

Welfare needs can also be classified in many other ways. For example, it is possible to make a distinction between those needs that are required throughout people's lives and those that are needed at particular periods. Thus, food and shelter are needed throughout life, whereas health care is needed at particular times. In addition, there are some sub-sections of the population with particular needs such as the blind, single-parent families, the infirm elderly or the ill. R. Rose (1986) argues that welfare needs typically exhibit a W-shaped distribution over time in accord with periods in the family life-cycle. Hence, there is a need for various types of help and care when one is young, when raising children, and when one is old. Some of these needs, such as those for food and housing, are tangible, whereas others, such as the need for social belonging, are more ambiguous and difficult to measure.

HOW CAN WELFARE NEEDS BE MET?

THE FAMILY

The numerous ways in which welfare needs can be met are all interrelated but it is perhaps best to begin with the family. Although in many countries welfare has been equated with provision by the state, it is important to remember that state provision of welfare needs is a relatively recent innovation. Throughout most of human history the main role of the state has been to provide defence of its territory against external attack and to maintain internal law and order (R. Rose, 1986). People have therefore looked to their families, relatives and friends to provide mutual support in times of need. Thus, the elderly were looked after by younger adult members of the household while children were cared for by their parents and siblings. It is not surprising, therefore, that the term economy is derived from the Greek word *oikonomia* which means 'management of the household' (R. Rose, 1986).

Throughout much of the twentieth century the bulk of work and caring in households has been undertaken by women while men have concentrated upon paid work in the external economy. However, before the industrial revolution, when the household was the key unit of production, women undertook a far wider range of tasks that were integrated with the work of men (such as spinning wool). Additional work undertaken by women such as cooking and tending animals meant that there was little time for the intensive childcare and cleaning that has characterised the role of many women in the advanced industrialised nations in the twentieth century. In the early days of the industrial revolution women became an important feature of the manufacturing workforce but gradually they were expelled from many industrial sectors and increasingly confined to the domestic sphere. In recent years these patterns have once again reversed with women becoming an ever increasing component of the workforce outside the household. However, this does not appear to have been associated with a major change in the overall distribution of domestic work amongst men and women.

Closely related to the use of family members for meeting welfare needs is the use of friends, relatives or other contacts, usually in the local neighbourhood of the household. In these situations, market exchange may take place but this need not involve money. For example, there may be some reciprocal exchange of baby-sitting as payment for tasks such as childcare. Because most of this activity is not officially recorded this is often known as the informal economy.

CHARITABLE AND VOLUNTARY ORGANISATIONS

The poor and destitute have long depended upon charity and handouts from the better-off in society. In the nineteenth century, following in the wake of the hardships produced by the industrial revolution, mutual benefit societies developed to ensure families against the risk of sickness, old age and death. At the end of the twentieth century what is known as the 'voluntary sector' is still an extremely important way of meeting welfare needs. This sector is extraordinarily diverse, including charitable trusts, pressure groups petitioning on behalf of particular sub-sections of society (such as the Child Poverty Action Group in Britain) and reciprocal self-help groups. Although often dependent upon charitable donations from individuals and companies, the voluntary sector also receives funds from the state, together with payments from users of their services. Relatively little is known about the voluntary sector compared with other parts of welfare systems but, as will be shown in Chapter 2, this sector has increased in importance in recent years following attempts to restructure the welfare state.

PRIVATE MARKETS

Throughout most of human history the provision of welfare within households and communities has taken place by exchange of mutual obligations and by barter. Much of this was swept aside by the industrial revolution which brought about a huge increase in the quantity of goods and services that could be produced. This in turn required increased use of monetary exchange and the development of private markets to allocate these products (sometimes termed *monetisation* or *commodification*). Thus, basic welfare goods and services such as shelter, food and health care became available through private markets. In the late twentieth century many welfare needs are satisfied by private markets. For example, private sector employers provide incomes, pensions, health insurance and education for their workers, and individuals and families can purchase education, health care and pensions from private sector institutions.

One of the major consequences of allocating welfare needs by private markets is that they tend to lead to inequalities in provision, since the poorest groups are often unable to afford goods and services. In addition, there are some people who might forgo provision for long-term needs, such as disability, poor health and old age, by satisfying short-term desires for the sorts of consumer goods that capitalist societies can provide in abundance. The state has therefore frequently adopted a paternal role through compulsory insurance and social programmes. These services, which are deemed so desirable that everyone should be entitled to them, are often termed *merit goods*. These are goods and services in which the benefits to the community as a whole exceed those to the individual. The consequence is that individuals will tend to consume too little of them for the common good. However, in practice what constitutes a merit good reflects not so much technical economic criteria as political decisions and cultural values.

THE STATE

As stressed above, the welfare state is a relatively recent feature of human societies and it achieved its zenith in the twenty to thirty years after the Second World War. However, it is important to realise that states have assumed responsibility for the poor and destitute for many hundreds of years. Most European nations had something similar to the English Poor Law which gave local authorities responsibility to administer relief to the destitute. Furthermore, despite the ideology of *laissez faire* that accompanied the growth of industrial capitalism in the nineteenth century, most states implemented various acts of legislation to regulate factory hours, improve the quality of housing and to deal with the problems affecting public health. Indeed, many of the foundations of the welfare state involving legislation covering industrial accidents, health and unemployment insurance, were laid down in the late nineteenth and early twentieth centuries. Nevertheless, the early Poor Law measures were concerned with control of vagrancy rather than the well-being of the poor. Similarly, the social legislation of the nineteenth century was strongly motivated by the desire to preserve public order. Furthermore, despite the very early legislation, it was only after the Second World War that welfare services developed rapidly to form what is today recognised as the welfare state.

The term welfare is often equated with the direct public provision of services such as health, housing and education on a non-market basis. However, there was a second and equally important element to the welfare state — a system of arrangements that ensured a relatively high standard of living for the majority of the population through policies that achieved full employment, relatively high minimum wages, safe working conditions and income transfers to deprived minorities. The past tense is used here because although states are still heavily involved in the direct provision of services, many governments, such as those in Britain in the 1980s, have reneged on this second set of obligations. The notion that states should have responsibility towards ensuring that all their citizens have a 'decent' standard of living has been termed *welfare statism* (Pfaller et al., 1991).

Some economists have argued that the state becomes involved in the provision of certain goods and services because these goods and services have properties that make it difficult if not impossible for them to be allocated by private markets. This interpretation is known as the Theory of Public Goods (Samuelson, 1954; Musgrave, 1958). In contrast to private consumption goods, such as clothing, food and housing, which can be consumed only by an individual or a household, public consumption goods have three important characteristics that prevent individual consumption. First, there is *joint supply*, which means that if a good can be supplied to one person, it can be supplied to all others at no extra cost. Second, there is *non-excludability*, which means that it is impossible to withhold a good from those who do not wish to pay for it. Third, there is *non-rejectability*, which means that once a good is supplied it must be consumed by all, even those who do not wish to do so. Although this theory has been influential, it is difficult to find clear support for it. The classic example that is used to illustrate the theory is defence: once an army is raised it is difficult for the average citizen to opt out of the protection provided. However, even the extent to which a country can be defended will vary across space. There are, in fact, relatively few technical reasons why a service should be provided by the public rather than the private sector. It is certainly difficult to provide some natural monopolies such as water, electricity and gas through private markets made up of numerous entrepreneurs, but these utilities can be provided through private companies if they are regulated monopolies. The decision whether to allocate goods and services through either public or private channels therefore reflects not so much the characteristics of the products being supplied but political decisions.

The influence of political decisions also undermines many of the other criteria that have been used to try and explain the nature of public sector involvement in welfare provision. For example, the distinction between goods and services is of relatively little use. It is certainly true that *collective consumption* is predominantly concerned with service activity, i.e. that which involves expenditure, and which does not involve the physical transformation of a product (as in the manufacture of a good). However, most so-called service activity involves the use of products in various forms, and the public sector is also concerned with the distribution of goods such as housing.

A more useful distinction is whether or not goods and services are allocated on market or non-market principles. One of the major reasons for the intervention of the public sector in the allocation of goods and services was dissatisfaction with the inequalities generated by a dependency upon market principles. However, it should also be recognised that many publicly owned services are traded on a commercial basis whilst the consumption of privately owned goods and services is bolstered by public subsidy. In the United Kingdom, for example, tax relief on mortgage interest amounts to a vast subsidy to home owners, many of whom are the more affluent members of society. Whilst non-market criteria can help to meet the needs of the least well-off, they can also be utilised by powerful groups who wish to derive concessions from the state.

Despite the diversity of ways in which welfare needs can be met, there are some similarities in arrangements in the advanced industrial nations. For example, despite the enormous expansion of state welfare services in the twentieth century, the vast bulk of caring is still undertaken within families. At the opposite extreme, the vast majority of educational provision is undertaken by the state. Indeed, education was one of the first responsibilities undertaken by states in the late nineteenth century, even in nations that were reluctant to engage in public intervention in the workings of private markets. This pattern suggests that an educated populace is crucially important for a modern industrialised nation and too important to be left to the uncertainties of the market. However, one should not pursue this argument too strongly. One might make the same argument for health, for example, yet this service shows an enormous variation in modes of provision between different countries. Indeed, the vast majority of services show wide variations in modes of provision, not only between different countries but also within their boundaries. It is because of this enormous variation that a geographical perspective is essential to understand the changing geography of the welfare state.

GEOGRAPHICAL VARIATIONS IN WELFARE STATES: THE INTERNATIONAL SCALE

One of the most remarkable features of the twentieth century has been the widespread growth of welfare expenditures in all the advanced capitalist economies. Thus, by the mid-1970s between a quarter and a third of the GNP of the major European countries was devoted to what can generally be termed social expenditures (Pierson, 1991). Similarly, in the United States, often regarded (somewhat simplistically) as a 'laggard' in welfare terms, over 20 per cent of GNP was being devoted towards the welfare policies in 1981. Even in Japan, where the welfare state is least well developed, over 16 per cent of GNP was spent on the social budget (Pierson, 1990). However, these aggregate figures conceal wide variations in the character of welfare states.

Many classification systems have been devised to account for this complexity of welfare states but most boil down to an assessment of the extent to which the state intervenes to affect market mechanisms. Table 1.1 is an attempt to summarise and integrate these classifications. Those typologies with underlying similarities are located close to each other. There is, however, no simple correspondence between the types of welfare states listed in the columns.

At the top of Table 1.1 is one of the classic and most cited typologies of welfare states – the threefold division made by Titmuss (1974). First, he distinguished the *residual* welfare model in which welfare institutions come into operation only as a last resort when the family or private markets fail to meet welfare needs in a satisfactory manner. Second, there is the *industrial achievement-performance* model in which social policy is geared towards the smooth functioning of the economy. Third, there is the *institutional redistributive* model in which universal services are allocated by the state on the basis of need. These are obviously ideal types and most countries will fall somewhere between these extremes, possibly incorporating elements of all three of these types.

Mishra (1984) makes a distinction that is similar to the last two categories postulated by Titmuss. First, he distinguishes the *differentiated* or *pluralist* welfare state in which social policy is distinctive from, and unrelated to, economic and industrial policy. This contrasts with a second type, the *integrated* or *corporatist* welfare state in which social welfare is seen as closely related to industrial sectors.

Therborn (1987) makes a similar categorisation of welfare states, not only on the basis of their degree of intervention in market mechanisms, but also on the degree to which they are committed to full employment. Some countries, such as Germany, have generous social entitlements but limited commitment to full employment, whereas in Sweden, social policy has a strong labour market element. In Japan, in contrast, there are limited social entitlements but a strong commitment to full employment. The 'market-oriented' welfare states of the English-speaking world – the US, Canada, Australia, New Zealand and the UK – would seem to have the worst of all worlds according to Therborn's typology, with both limited social rights and limited commitment to full employment.

The final classification scheme is derived from Esping-Anderson's (1990) highly influential analysis of *welfare regimes*. He used seven variables to analyse

Table 1.1 Typologies of the welfare state

TITMUS (1974)

Residual welfare model	Industrial achievement-performance model	Institutional redistributive model
– state is a temporary last resort	– welfare institutions are an adjunct of the economy	– universal services allocated on the basis of need

MISHRA (1974)

	Integrated or corporatist welfare state	Differentiated or pluralist welfare state
	– social sector integrated into economic and industrial policy (Austria)	– social welfare sector is distinctive and unrelated to economic policy (UK)

THERBORN (1987)

Market-oriented welfare states		Strong interventionist welfare states
– limited social rights, low commitment to full employment (Australia, Canada, US, UK, New Zealand)		– extensive social policy, strong commitment to full employment (Sweden, Austria, Norway)
	Full employment-oriented welfare states – low social entitlements, commitment to full employment (Japan)	Soft compensatory welfare states – generous social entitlements, low commitment to full employment (Belgium, Denmark, Netherlands France, Germany, Ireland, Italy)

ESPING-ANDERSON (1990)

Liberal welfare state	Conservative/corporatist welfare state	Social democratic welfare state
– dominated by market, modest benefits, means testing (US, Canada, Australia)	– strong state welfare orientation, minimal private insurance, conservative attitude towards family (Austria, France, Germany, Italy)	– state is principle means of realising social rights, graduated universal insurance system, commitment to full employment (Sweden, Norway)

---

the structure of welfare provision in eighteen of the advanced industrial nations. Esping-Anderson distinguishes between three different types of regime of which, it is argued, countries display elements in varying degrees (see also Table 1.2). The *liberal* welfare state is dominated by the logic of the market and encourages

Table 1.2 Esping-Anderson's classification of welfare states

Index score	Conservatism	Liberalism	Socialism
High	Austria	Australia	Denmark
	Belgium	Canada	Finland
	France	Japan	Netherlands
	Germany	Switzerland	Norway
	Italy	United States	Sweden
Medium	Finland	Denmark	Australia
	Ireland	France	Belgium
	Japan	Germany	Canada
	Netherlands	Italy	Germany
	Norway	Netherlands	New Zealand
		United Kingdom	Switzerland
			United Kingdom
Low	Australia	Austria	Austria
	Canada	Belgium	France
	Denmark	Finland	Ireland
	New Zealand	Ireland	Italy
	Sweden	New Zealand	Japan
	Switzerland	Norway	United States
	United Kingdom	Sweden	
	United States		

Variables used in index:

Conservatism: number of pension schemes available; amount of expenditure on government employees' pensions

Liberalism: proportion of state benefits subject to means testing; importance of private sector in pensions

Socialism: proportion of population entitled to benefits; degree of equality in benefit provision

Source: Esping-Anderson (1990)

the private provision of welfare in the form of private insurance schemes and occupational-based welfare. Such welfare states have a strong work ethic and there are usually limited welfare benefits for the most deprived minorities. The US, Canada and Australia are cited as examples of welfare states with strong elements of liberalism. The second type of regime – the *conservative/corporatist* welfare state – is less concerned with liberal conceptions of market efficiency and has a strong commitment to state provision; correspondingly, private insurance and occupational benefits are minimal. Examples of countries with strong elements of this welfare regime are Austria, France, Germany and Italy. The influence of the church in these continental European countries means that social policy directed towards the family has a conservative bias. Finally, there is the *social democratic* welfare state, characterised by state intervention to usurp the market. Universal participation in earnings-related benefits are encouraged by a commitment to full employment. The Scandinavian countries have strong elements of this type of welfare regime.

Despite these many differences in the character of welfare states, it is possible to distinguish between three main economic blocks in the advanced capitalist nations with differing types of approach to satisfying welfare needs: Europe, the United States and Japan. The differences in the ways in which welfare needs are satisfied in these blocks depends to a large extent upon varying uses of the state, private markets and the family.

EUROPE

It is appropriate to begin a consideration of welfare states with western Europe since it is in this block of countries that the idea of the welfare state and *welfare statism* is most developed. In Europe, in general, the state is regarded as having a major responsibility for caring for its citizens through income maintenance and the provision of welfare services such as health and education. In addition, the state is expected to be an active participant in directing and organising the economy.

As suggested by the classifications schemes outlined in the previous section, there are wide variations in the nature of welfare states within western Europe, but they share a number of similarities compared with the other major economic blocks. The Scandinavian countries are the most extreme examples of the European approach since they have a strong sense of collective solidarity and welfare is seen as a collective responsibility that is undertaken by the state on behalf of its citizens. The welfare state is often seen as a means of ensuring the highest rather than the lowest-quality standards, but rights and benefits are often less extensive than in Germany, France and Italy. The UK has the welfare statism to be found in these continental European countries but in recent years has moved further towards the welfare system to be found in the US, which is dominated by the ethos of the market.

There are other differences in the character of European welfare states. Until the last decade, the UK was somewhat unusual in the extent to which it depended upon both state financed and state provided welfare services, most notably in the context of housing and health care. In contrast, most continental European nations depend to a much greater degree upon private and voluntary provision of welfare services, even if they are funded by some national insurance scheme. Furthermore, it is also important to remember that the continental European welfare states have, for the reasons to be elaborated below, been under increasing pressure in recent years to reform their generous social provisions.

THE UNITED STATES

The US is widely acknowledged to be a world leader in industrial, financial and technological spheres as well as being a key innovator of democratic institutions. However, as a welfare state, the US has often been seen as underdeveloped and limited in scope. In contrast to the continental European emphasis upon social solidarity and the collective responsibility of society to care for the less fortunate, social relationships in the US tend to be seen in individual terms. Thus, the individual, supplemented by the family and local community groups, is regarded as having the primary responsibility for personal welfare, with the state intervening only in the last resort. Consequently, the US is dependent upon a complex web of private and voluntary agencies – both profit and non-profit making – that help to meet welfare needs.

A number of factors have been suggested to account for the limited development of the welfare state in the US. Many point to the pioneering spirit of individualism that was necessary to develop the frontier which, it is claimed, has fostered a widespread hostility towards state intervention. In addition, the fragmentation of US society into numerous racial and ethnic groups is widely seen as having prevented the emergence of a unified labour movement and left-wing political parties of the type that were important in bringing about the development of the welfare state in Europe. Finally, there is in the US a strong tradition of federalism that has inhibited the evolution of a national approach to social problems.

However, it would be wrong to regard the US as being devoid of state intervention for, although somewhat later than in Europe, there were two periods of rapid innovation in the welfare sphere. The first period was the New Deal of the 1930s in response to the hardship of the Depression. Under the presidency of Franklin D. Roosevelt, a national system of protection against unemployment, old age, death and disablement was established. The second period of innovation was the Great Society Initiative which began in the 1960s under President Johnson. Together with the expansion of policies that could benefit everyone, such as social security (insurance for old age, disablement and death) and the introduction of Medicare (contributory health insurance for the aged), various programmes were introduced for the poorest minorities including Medicaid (non-contributory health care for the poor) and AFDC (Aid to Families with Dependent Children). This so-called War-on-Poverty continued under presidents Nixon and Ford, eventually running out of steam under President Carter's administration. Although levels of social expenditure never reached European levels, the costs of these programmes escalated at a time of increasing economic difficulty. In addition, there was widespread disillusionment with many of these programmes as they seemed incapable of halting the escalating problems of crime, poverty and family breakdown. In general, there is in the US a strong preference for programmes that favour those who are regarded as self-reliant and independent. Thus, recent attacks upon welfare in the US have been upon residual programmes for the poor rather than upon policies such as those in the sphere of pensions and education which affect a larger proportion of the population.

JAPAN

Since the Second World War the growth of the Japanese economy has been phenomenal. In the 1950s and 1960s the Japanese economy grew in real terms at about 10 per cent per annum and in recent years the growth rate has been twice that of other OECD (Organization for Economic Co-operation and Development) nations. Consequently, Japan's economy is now huge, outstripping that of the UK, Germany and France combined. Japan now has one of the highest per capita incomes in the world, one of the lowest rates of infant mortality, and the second highest proportion of students participating in higher education (the US having the highest proportion).

However, the structure of the welfare system in Japan is very different to that in other nations of the OECD. In Europe and the US family links are declining as divorce rates rise and the numbers of single-parent families increases. Consequently, many of the tasks that were previously undertaken by different generations in the extended family have been taken over by the state or other non-family sectors of the welfare system. In Japan, however, there are still strong links between the generations; for example, most of the elderly try to live with one or more of their children and help to care for their grandchildren. Correspondingly, the amount of welfare provision from the state is much less than in typical European countries. This limited public sector in Japan reflects the fact that the welfare state came much later to this country than to European countries, only really developing after the Second World War. In fact, as in the US, the main component of welfare provision in Japan is state provided pensions. Indeed, the rapid increase in the numbers of elderly persons in Japan has been one of the main reasons why spending on social security has risen so rapidly in this country in recent years.

In Japan there is a strong emphasis upon group solidarity which affects all aspects of life including industrial relations and government. However, unlike the UK, this collectivist ethic is expressed through companies rather than through statutory social services. Thus, the major corporations play an important welfare role in Japan, providing housing, family allowances and social facilities for their workers. However, there are considerable variations in the value of the benefits from different companies – the major corporations tending to be more generous than smaller companies. The system of lifetime employment means that the age of retirement tends to be relatively early in Japan. Although in many of the larger companies retired employees are allowed to go on using certain services such as health care and holiday homes, and some of them even find a second job in one of the smaller associated companies, most retired workers initially find that they need additional private health insurance and housing (Pinker, 1986). Use is also made of the highly diverse voluntary sector until statutory services and state pensions can be used at the age of 70.

These arrangements contrast markedly with the pervasive ethic of individualism to be found in the US. However, the US and Japan are similar in not adopting the assumptions about collective action by the state that are held in Europe. In the latter the state is expected to be an active participant in directing the economy as well as providing welfare for all its citizens. In Japan, in contrast, the family, together with large employers, substitute for the state in providing welfare. Thus, Rose and Shiratori claim:

Many Americans accustomed to think of Japan as alien may nonetheless find the dominant political values of the right-of-centre Liberal Democratic party more familiar than the left-of-centre values often dominant in Christian democratic Europe. An ideological map would locate Japan as the westernmost of the 'Sun Belt' states, to the right of Texas or California, and a few centres of ideas on America's Atlantic Coast might even appear closer to Europe than California . . . A

Japanese politician would find himself ideologically more at home in Texas than Sweden or Germany, and a Texas politician would be more at home in Kobe than Copenhagen.

(Rose and Shiratori, 1986, pp. 6–7)

Because of the differing ways in which welfare needs can be met, there is no necessary correspondence between the level of state spending upon welfare services and the general health of the population. Thus, Japan spends relatively little of its GDP on social expenditures but has a higher rate of life expectancy – a critical indicator of welfare – than the US, Britain and most of Europe. Nevertheless, each of these differing ways of meeting welfare needs has different distributional consequences. For example, if the family is the main source of welfare, then single persons, widowed and divorced persons will be disadvantaged. If, on the other hand, the market is the primary mechanism for resource allocation, then those on lower incomes will tend to be disadvantaged. If the state is the main determinant of welfare, the access will depend upon the regulations and allocative criteria devised by state bureaucracies.

GEOGRAPHICAL VARIATIONS IN WELFARE WITHIN STATES

There are also substantial variations in the character of welfare systems *within* states. This is especially the case in countries that are socially and ethnically diverse and that have complex federal rather than unitary constitutions.

One of the main reasons for this variation within countries is because of what is often known as *jurisdictional partitioning*. Most nations find it necessary to devolve the administration of services down to units that are smaller than the nation state. This devolution is often done in the name of local democracy, the assumption being that a centralised government may be remote from the wishes of the public. Hence, these decentralised administrative units often have elected representatives with varying degrees of political autonomy. Other units are various types of special administrative district with varying degrees of political and consumer representation. Whatever the causes, one of the main consequences of this devolution is that these political and administrative units often vary enormously in the quantity and quality of goods and services that they provide. Typically, these variations reflect political ideologies and mirror some of the variations noted at the state level in the previous section. For example, left-wing dominated political units are more likely to be committed to welfare provision through state services whereas right-wing dominated units are more likely to rely upon market mechanisms, the voluntary sector and self-help. As at the state level, we also find a wide range of variations between these extremes. And, once again, as at the level of the nation state, there are a wide range of factors in addition to political ideologies that can affect the level of welfare provision, such as the level of economic development and local cultural traditions (see Pinch, 1985, ch. 2 for further details). A recent major shift in the structure of welfare states is the devolution of responsibility for services down to smaller administrative units. This has led to geographical variations in the nature of welfare reform. As will be revealed in subsequent chapters, research is beginning to reveal the ways in different local organisational cultures are crucial in affecting the speed and nature of change.

There are often considerable geographical variations in public service provision *within* these smaller administrative and political boundaries. One of the main reasons for this unequal allocation of services is the existence of distance-decay effects (sometimes also known as *tapering* effects). Even if public services are free at the point of supply, many services, such as schools, have to be located at specific points. Those who live further away will incur greater costs in getting to these facilities and are therefore likely to consume them less. Even if the service visits the consumer, geographical inequalities can result; for example, it will take an emergency service longer to respond to a distant location. In some instances, however, consumers wish to be distant from what are known as noxious facilities such as refuse disposal tips. As a consequence of these processes, there is within political jurisdictions and administrative boundaries a continual struggle on the part of consumers to minimise their distance from desirable facilities and to maximise their distance from undesirable services. There is a large body of literature that has examined the conflicts and political struggles that arise over the allocation of these resources (see Pinch, 1985 for a review). As will be shown in subsequent chapters, however, the recent restructuring of welfare states has radically altered many of the processes of service allocation and the ensuing conflicts. Inevitably, geographers have tended to look at spatial variations in the direct provision of welfare services such as schools, hospitals, day-care centres and social services rather than in the income maintenance schemes of the welfare state, since the latter tend to be geographically invariant throughout nations (even if they have different purchasing power because of variations in costs). However, given the growing importance of private occupational pensions, the latter have also begun to attract some attention from geographers (Clark, 1990a; 1990b).

This book will attempt to demonstrate that geography is not simply a manifestation of these differences in the character of welfare states; space – like time (which until recently has received the bulk of attention) – is a crucial element in the process of change.

PRESSURES FOR CHANGE IN THE WELFARE STATE

The aim of this section is to examine briefly some of the pressures upon the welfare state in recent years. A key element in producing change has been the conflict between, on the one hand, rising demands and expectations, and, on the other hand, falling resources. Each of these elements is now considered in turn (for a summary see Table 1.3).

Table 1.3 Pressures upon welfare states

Demographic trends

Increasing numbers of dependent groups (e.g. the unemployed, homeless, old people, the infirm elderly, single-parent families, disabled)

Growth of the 'new poverty' through polarisation of incomes

Feminisation of workforce (e.g. growth of low-paid jobs and increased need for day-care)

Cultural Factors

Rising expectations and desire for higher standards

Desire for diversity of provision

Strength of pressure groups (e.g. on behalf of disabled, aged, long-term sick)

Growth of consumer culture

Increasing cultural pluralism and diversity

Growth of 'new age' movements (e.g. alternative and complementary medicines)

Tax revolts

Intellectual/ideological developments

Growth (rebirth) of New Right ideas

Decline of faith in collective solutions to social problems

Economic trends

Increasing proportion of the population dependent upon a smaller workforce

Inelastic sources of local revenue, 'stagflation' (high unemployment and high inflation in 1970s)

Increased competition between nations for mobile capital combined with enhanced sensitivity of national economies to capital flows

Desire of transnational corporations for low social overheads

---

RISING DEMANDS AND EXPECTATIONS

As noted above, there has, until recently, been a steady increase in the proportion of the GDP of the advanced industrial nations devoted to welfare services. This gradual increase was the result of a number of forces. One of the most important forces has been demographic pressures, for these have increased the size of the client groups for many welfare services. Of particular significance has been the increase in the absolute numbers and proportions of the total population of persons over retirement age (see Table 1.4). The increase in the very elderly population aged 75 and above has put the greatest pressures upon social services, for it is this section of the elderly who experience the greatest incidence of sickness and disability. This growth in the proportions of the very elderly has led to increased demands for community-based forms of care such as home nurses, meals-on-wheels and health visitors. However, the problem of old age should not be exaggerated. In the UK, for example, the proportion of the population over 65 is currently about 16 per cent and is set to continue at this level until the 2020s and 2030s, when the 'baby boom' generation born in the 1950s and 1960s reaches retirement age.

Changing family structures, including increased rates of family breakdown and divorce, have also increased the need for welfare services. However, once again, the importance of these demographic and cultural factors should not be exaggerated. Indeed, some of these factors have been distorted into a 'crisis' by those on the Right who are anxious to curb the welfare state. By far the

Table 1.4 The elderly population as a percentage of the total population of the UK between 1981 and 2051

	Age group		
	60–74	75+	Total 60+
1981	14.4	5.8	20.2
1991	13.7	7.0	20.7
2001	12.8	7.4	20.2
2011	14.8	7.4	22.2
2021	16.1	7.9	24.0
2031	17.8	9.1	26.9
2041	15.4	10.6	25.9
2051	15.2	10.3	25.5

Sources: adapted from OPCS projections and Benington and Taylor (1992)

most important drain on the welfare state in years has been the increase in the social security budget in response to rising unemployment levels (and especially the increase in the long-term unemployed). We have also seen the rise of the so-called 'new-poverty' – relatively large sections of the population on very low incomes. This is related to increased flexibility in the labour market, the associated polarisation of incomes and a greater proportion of the population in need of income-support schemes (Benington and Taylor, 1993).

The costs of welfare services have also been increased by the desire for higher standards. This is manifest in improved schools, hospitals, libraries and sports facilities. These increasing standards are the product of numerous forces. In part, they reflect an era when there was a concern for rational, centralised planning. Research departments within both central and local government and the National Health Service (NHS) provided comparative indices of service provision and acted as a spur to improve standards. These research departments also unearthed information on the specialised needs of particular groups such as the disabled and the mentally handicapped. Professional bodies have also increased pressures for better standards. However, paradoxically, it is only in recent years with attempts to reign in spending that the quest for monitoring standards and assessing performance has begun to take central place in the operation of public services (see Chapter 3).

Expenditures on welfare services have also increased because of rising public expectations. In part, these rising expectations are a response to increasing service provision. Research has suggested that once a service is provided by the public sector, demand tends to increase; conversely, restricting information about services can lead to a dampening down of demand (Pinch, 1985). Indeed, it has been argued that state involvement in provision tends to increase demand for services; instead of the 'hidden hand' of the private market, there is a political body to whom representations can be made in the face of limited services. Demands for services have also increased because of the activities of numerous pressure groups. Some of these pressure groups are territorially based, attempting to preserve their own bit of 'turf' from invasion by undesirable activities. Alternatively, other groups may be attempting to gain extra facilities for their neighbourhood. Some pressure groups may be petitioning for sections of the population with particular needs. Numerous factors have been suggested to account for the increase in these pressure groups including: the growth of the women's movement; increasing levels of education; an increasing awareness of ecological issues together with the growth of various ecological movements; increasing levels of home ownership; and the development of consumerism.

DECLINING REVENUES

Coincident with these increasing demands upon the welfare state have been declining revenues. In large measure these are a consequence of the economic problems encountered by the industrialised nations in the 1970s. For a variety of reasons, a combination of high inflation and rising unemployment led to the need for expenditure cuts in welfare services. The extent of the pressures upon the welfare state in particular countries can be related to some extent to the nature of their welfare systems (Walsh, 1995). In countries where welfare issues have tended to be divorced from issues of economic growth – the UK, Australia and New Zealand – then public spending was seen as a major contributor to economic decline. However, in those nations with more integrated welfare states, such as Sweden and Germany, economic problems impacted to a lesser degree upon the welfare state.

A crucial element in the restructuring of welfare systems has been globalisation – the tendency for economic and cultural developments to be increasingly integrated on a global scale. A prime force behind globalisation has been the major transnational corporations. Given their capacity to shift production throughout much of the world, they are engaged in a continual process of comparing countries to find the optimal locations for their activities. This means that countries have to provide a number of attributes if they are to be attractive to transnational capital. One of these inducements is low corporate taxes and social overheads in the form of a small welfare state.

CHANGING IDEOLOGIES

Walsh (1995) stresses the fact that public services are changing throughout the world. Thus, substantial changes have been made even in countries governed by social democratic political parties such as in Sweden, Australia, Germany and New Zealand. At times, these changes have been introduced reluctantly, suggesting that reform of the welfare state is not simply the result of political ideology but is often a pragmatic response to the pressures noted above. Nevertheless, it is undoubtedly true that the pace of change has been greater in countries dominated by right-wing political parties, such as the UK. Furthermore, however pragmatic may be the responses, pressures do not automatically produce solutions; they need to be perceived and acted upon by people. New ideas have therefore been important in reshaping the welfare state. In particular, there has been a barrage of criticism directed against the welfare state in the last fifteen years by those who have been termed the New Right.

Although there are many variations in the ideas of those who comprise the New Right, they tend to share in common a belief in the value of market mechanisms as the most efficient ways of ensuring the production and distribution of goods and services (see Dunleavy, 1991 for a review). It is argued that lacking the discipline of market signals, public sector organisations are inefficient, hence they tend to be staffed by large self-serving bureaucracies who are insensitive to the needs of their client groups. Furthermore, some argue that politicians are relatively powerless to influence the public sector. Indeed, one line of reasoning argues that politicians encourage public spending, since in their attempts to win office they tend to raise expectations that cannot be met.

There are many different strands to neo-conservative thought. In one extreme form, *libertarians* feel that governments have no right at all to get involved in the actions of individuals and instead should adopt a 'night watchman' role, protecting the property rights of individuals (Nozick, 1974). However, such views – essentially permitting people to do what they like provided they do not interfere with others – are opposed by a strong moralising tendency amongst some on the Right who believe in bolstering a wide range of moral values, often centred around a traditional notion of the nuclear family. We have also seen the growth of the 'new citizenship' theory which stresses the duties and obligations of individuals rather than their inherent rights to universal benefits from the state (for a critique see Smith, 1989). Other scholars, less extreme than the libertarians, argue that the state has become too large, imposing a drag upon the private sector. This led in the early 1980s to policies aimed at restricting the size of public spending.

One argument is that the welfare state is an expensive luxury that the advanced industrialised nations cannot afford if they are to compete successfully with countries whose industries are not encumbered by high rates of tax, high wages and a host of rules and regulations surrounding their activities of business. A second, stronger view is that welfare states are not simply an expensive luxury but a major obstacle to increasing competitiveness. An integral part of this second argument is the notion that 'generous' welfare payments to the unemployed serve to undermine the work ethic and discourage people from seeking employment. Indeed, many neo-conservatives ascribe the decline of the conventional nuclear family to the way in which the welfare state has assisted alternatives.

Although the New Right have set the pace in criticising the welfare state, there was also a loss of confidence amongst 'mainstream' supporters of the welfare state. There was a consensus about the value of welfare states in the 1950s and 1960s but when this became weakened by economic and social difficulties in the 1970s, those on the Right entered to fill the vacuum. Much of the loss of confidence in the welfare state centred around the obvious failure of Keynesian demand management of the economy. This involved government spending in times of recession to boost the capitalist economies, but the value of this was destroyed by the 'stagflation' of the 1970s – a combination of high unemployment and high inflation. In addition, the rediscovery of poverty in the 1960s and the failure of many social programmes such as urban redevelopment also undermined faith in *social engineering* – the capacity of society to produce comprehensive rational solutions to social problems.

The welfare state has also been criticised from those on the Left. Although there are many different types of interpretation of the welfare state (see Pinch 1985 and Johnston, 1993 for a review of these theories), they tend to share the view that the fiscal crisis of the state has arisen because of the contradictory demands of capitalism. On the one hand, the state is required to undertake activities to legitimate the system but, on the other hand, it is also needed to socialise many of the costs of production to bolster profits. Corporations are unwilling to carry the burden of these costs, hence the so-called 'crisis' of the welfare state.

CONCLUSION

This first chapter has outlined the basic elements of welfare systems and has indicated how these elements vary geographically, both between and within the boundaries of nation states. This chapter has also outlined some of the pressures inducing change in welfare structures in recent years. These pressures have meant that the relationships between the key elements of welfare states have been in a continual state of change. In particular, there has been a reduction in levels of direct state provision and a greater reliance placed upon provision by families, the voluntary sector and through markets. In addition, there have been substantial changes in the management of those elements that remain funded by the public sector. As will be shown in the following two chapters, these changes show considerable variations between places.

FURTHER READING

There are literally hundreds of books on welfare states. The following are just a few of the useful introductions to recent changes in welfare structures: Cochrane and Clarke (1993), Glennerster (1995), Mishra (1984), Pierson (1991) and Williams (1989). It is also worth seeking out Esping-Anderson's (1990) influential work. Johnston (1993) provides an extremely clear summary of the pressures upon welfare states. The classic New Right text is Hayek (1956). Offe (1984) is a good example of a recent neo-Marxian interpretation of the problems of the welfare state.

Introductions to geographical perspectives on welfare services include Bennett (1980), Curtis (1989), Duncan and Goodwin (1988), Jones and Moon (1987), Joseph and Phillips (1984), Painter (1995) and Pinch (1985). A comprehensive review of issues of 'territorial justice' is to be found in Boyne and Powell (1991).

CHANGING THE WELFARE STATE: 'CHIPPING AWAY AT THE EDGES'

This chapter examines some of the recent changes in the structure of the welfare state in response to the pressures noted in Chapter 1, together with some of the geographical manifestations of these changes. The most basic change in welfare regimes throughout the world has been a decline in direct provision by the state. This decline has put greater reliance upon other sources of welfare provision: by families, by charities and by private markets. Dependence upon a variety of modes of welfare provision has been termed *welfare pluralism* (Johnson, 1987) and the *mixed economy of welfare* (Pinker, 1992). In addition, there has been an erosion of many of the broad features of welfare statism: the commitment to full employment, workers' rights and adequate minimum standards for all citizens (Pfaller et al., 1991).

RATIONALISATION

One of the first changes in the recent bout of welfare restructuring has been a reduction in the scale of services provided directly by the public sector. In the UK in the early 1980s these reductions became known as 'the cuts'. Drawing upon the theory of industrial restructuring and the closure of manufacturing plants in the private sector, we can also refer to this type of change as *rationalisation* – the closure of capacity (Massey, 1984; Pinch, 1989). There are many examples of reductions in spending and provision levels in the context of public services including schools, hospitals and specialist welfare centres.

The first western government that was elected on a platform of reducing taxes and curbing the size of government was the Conservative Administration headed by Margaret Thatcher that came to power in the UK in 1979. However, it proved difficult for this first Thatcher Government to reduce taxes because of its initial reliance upon a monetarist policy of squeezing inflation by reducing the money supply. This policy led to massive bankruptcies, deindustrialisation and a huge increase in unemployment which required a huge increase in social security and welfare payments. Furthermore, it has proved difficult to reduce many public services in a radical way although there have been substantial reductions in real terms in many services. The net result has been a redistribution of taxes away from direct to indirect forms of taxation rather than an overall reduction. Taken overall, the Thatcher Administrations amounted to a policy of gradual attrition of the welfare state rather than radical dismemberment. However, a key component of the attacks were ideological, attempting to cut away support for collectivist solutions to social problems in an attempt to prepare the ground for more radical solutions at some future date.

Eighteen months after the first Thatcher Administration, the first Reagan Administration was elected in the US, also committed to neo-conservative policies. There were fewer obstacles to welfare reform in the US. To begin with, as noted in Chapter 1, there was a smaller welfare system in the US and less opposition to change. However, opinion polls both in Britain and the US revealed that in neither country was the majority in favour of major reductions in education, health care and income maintenance programmes. Hence, in the US, programmes that benefit the broad middle class, such as education and pensions, were left intact whilst social programmes for the poorest were cut. The latter included eligibility criteria for Aid for Families with Dependent Children (AFDC), Medicaid and food stamps.

GENDER AND 'THE CUTS'

A key issue for many researchers has been the extent to which cuts in public expenditure have had a differential effect upon men and women. There is now a substantial body of evidence that indicates that women have been especially disadvantaged by recent changes in the welfare state (for a review see McDowell, 1989). This unequal treatment seems to have arisen for two main reasons: first, because women comprise the main recipients of welfare services that have been cut; and second, because women also comprise the bulk of the producers of these services that have been reduced (Webster, 1985).

To illustrate, the low rates of pay and lack of job security experienced by many women means that they are often more likely to be dependent upon welfare benefits of various types compared to men. Given that women often bear the main responsibility for looking after children, cuts to childcare services have had a particularly detrimental effect upon the access of mothers to employment opportunities. In addition, demographic trends mean that women are greater consumers of welfare services. On average, women live longer than men and therefore comprise a greater proportion of those requiring long-term care such as nursing homes, home helps, meals-on-wheels and community services. In addition, women dominate many of the occupations to be found within the welfare state such as nursing, teaching and social work. Thus, women constitute 75 per cent of the manual labour force in local government and are the same proportion of the total health service (Bakshi et al., 1995). Reductions in service levels have therefore had a particularly big impact upon the employment prospects of women.

In general, it would seem that poorer, marginalised groups have suffered most from policies of fiscal retrenchment. For example, there is evidence that various minority groups suffered disproportionately from the fiscal crisis in New York in the mid-1970s (Sheftner, 1980). Le Grand (1984) suggests that the class composition of both the consumers and producers of services is important in understanding the differential impact of cuts. When there are powerful middle-class producers and consumers, such as in the case of education or the British NHS, it is politically dangerous to make large cuts in expenditure. Thus, as Mohan (1989) notes, the first Thatcher Administration withdrew from radical change of the NHS after considering the damaging electoral consequences that would ensue. Social work is rather more vulnerable, however, because although it has a middle-class producer group, this occupation is generally unpopular and the recipients tend to be the poor. Most vulnerable of all, however, is a service such as refuse collection which has working-class producers. Although Le Grand's explanation is compelling, as indicated above, there would also seem to be an important gender element at work. Thus, in the NHS it was ancillary services dominated by women (often of minority status) that were the first to be subject to policies of contracting-out (see Chapters 3 and 5).

The geographical implications of service reductions have been most often studied in the context of education (Bondi, 1988). Although many of the criteria involved in school closures are apparently rational and technical in character, involving factors such as class size and enrolment levels, in practice, the outcomes often impact most severely upon poor communities that are less able to organise to resist closures. As with all social processes, of course, there is a wide range of contingent factors that can serve to influence and cause deviations from general trends.

PRIVATISATION 1:
ASSET SALES AND RESIDUALISATION

Hand-in-hand with a reduction in the scope of public services has been the so-called privatisation of services. However, privatisation is a term that has attained a high level of use and a low level of meaning, for there are a wide variety of processes that can be encapsulated by the term. One effective way of grasping the concept of privatisation is to consider variations along two dimensions, as shown in Table 2.1. On one axis we have sources of funding, by either the state or the market, and along the other axis we have the ownership of the service, again whether in public or private hands. The publicly owned, state funded services are those that comprise the core of the welfare state. The most simple form of privatisation, therefore, is when public sector assets are sold to the private sector. The most common form of *asset sale* in the UK in recent years has involved the nationalised utilities of gas, electricity and water supply. These services have welfare implications through their employment levels and the charges they make to consumers but do not have a direct welfare role. Since many welfare services are intended for the most disadvantaged in society who lack purchasing power, the extent to which welfare services can be provided by private markets is extremely limited.

Table 2.1 Forms of privatisation

	Private ownership	Public ownership
Market Funding	Sale of assets	Commercialisation, Corporatisation
State Funding	Contracting-out	Public provision

Source: adapted from Stubbs and Barnett (1992)

ASSETS SALES: PUBLIC SECTOR HOUSING

The most important example of asset sales of the welfare state in the UK has been the sale of public sector housing to tenants (Crook, 1986; Dunn et al., 1987; Forest and Murie, 1986; 1991). It could be argued that there is no overall reduction in service levels under this policy, since the tenants usually still occupy the accommodation that they purchase. However, once the properties are sold, the local authorities no longer have the right to nominate families to these properties and this has important implications for the role of the public sector in the sphere of housing. This section examines some of the background to this important policy.

One of the unique features of the British welfare state is the scale of housing that is directly owned, subsidised and administered by the state (through local government). It might be thought that reductions in the scale of public sector housing in the UK in recent years have been a response to the lack of growth in the British economy. For example, one of the conditions of the financial support Britain received from the International Monetary Fund (IMF) in 1976 was the imposition of reductions in public spending. Hence, the Labour Government of that time responded by cutting the subsidies to local authority housing. However, the scale of reductions in spending in recent years have far exceeded those that would be dictated by fiscal austerity. For example, whereas the Labour administrations of the 1970s reduced public expenditure on housing by 18 per cent, between 1979 and 1989 the following Conservative governments reduced housing expenditure by no less than 64 per cent (Hamnett, 1989). Indeed, in the early days of 'the cuts', housing accounted for the vast majority of the reductions in public welfare spending.

The scale of these reductions illustrates a crucial feature of public sector housing in the UK – namely that it has been the source of a crucial ideological battle. It is certainly true that at times during the 1950s there was something of a consensus about the merits of social housing and, under the Conservative administrations headed by Macmillan, the construction of local authority housing reached some of the highest ever levels in the UK. However, following the drift towards the Right under the Thatcher Administrations of the 1980s, not only was owner occupation celebrated but there was a direct attack on the scale of local authority housing.

The main weapon in this attack was the right-to-buy legislation which enabled existing tenants of local authority housing to purchase their properties at discounts of up to 70 per cent from their estimated market value. Prior to this legislation, sales were at the discretion of the local authorities and, since many were hostile to the policy, sales had been extremely small. However, following the right-to-buy legislation, sales boomed with a peak of 200,000 properties sold in 1982.

The significance of this policy should not be underestimated. For example, between 1979 and 1983 the policy generated £2 million in receipts – a sum exceeded only by the sale of British Telecom. However, there is now a substantial body of research that reveals that the sale of local authority housing has been uneven amongst social groups, between different types of housing and between different parts of the country. To begin with, those households who have purchased local authority properties have tended, on average, to be older than tenants, often at the stage of the life-cycle in which children have either left home or are at work and therefore bringing additional income into the household. In general, these tend to be multiple-income households in which both the male and female partners are in work. In addition, those in non-manual occupations have been more likely to buy their properties compared with those in manual occupations. Conversely, single-person households, single parents, widowed or divorced persons, households in which the head is unemployed, and retired households have been less likely to purchase their properties.

There have also been wide variations in the types of property that have been purchased. Although at the beginning of the 1980s flats constituted a third of the total public sector housing stock, in most of the years since the introduction of right-to-buy legislation less than 4 per cent of sales were in this category. The most popular dwellings for purchasers were three-bedroom semi-detached family dwellings. However, it is not just the character of the dwellings or their inhabitants that have been crucial in determining the scale and character of sales; the role of local authorities has also been important. Labour controlled local authorities have often been hostile to the sale of their housing stock and, in the early days at least, attempted to impede this policy. Conservative controlled authorities, in contrast, have tended to be much more enthusiastic about sales (Hoggart, 1985).

Putting these factors together has led to some interesting geographical patterns. Within cities, it has tended to be the higher quality, suburban estates, with three-bedroom family housing in areas with relatively low rates of unemployment and good reputations, in which sales have been most prevalent. Conversely, it is local authority estates that have acquired poor reputations, often associated with high levels of unemployment, crime and single parenthood, that have tended to have low rates of sales whether they be in suburban or inner-city areas. Between local authorities, the patterns are less immediately obvious. However, a close inspection reveals that sales have been greatest in free-standing cities and smaller towns outside of London and the major con-urbations.

However, the geographical distribution of council house sales is not just a reflection of the social characteristics of the inhabitants, the physical character-istics of the properties or the political complexion of the local council. Thus, when these factors have been introduced as variables in quantitative studies, there is usually a good deal of variation that still remains to be explained (Kleinman and Whitehead, 1987). One reason for this is that there are local factors at work in particular areas. For example, the overall scale of the public sector in an area seems to be an important factor affecting the rate of sales. Thus, if the scale of local authority housing is a relatively small proportion of the total housing stock in an area then, irrespective of the individual and political factors noted above, there will be an added incentive to purchase a property. Conversely, if there is relatively little owner occupation in an area there may be a disincentive effect to purchase a property. A related element here seems to be the economic buoyancy of the local area. It appears that sales are higher in economically prosperous areas; presumably escalating house prices in such areas in the 1980s encouraged households to take advantage of the capital gains on offer.

The net outcome of these changes is that public housing in the UK has ceased to become available to meet the general housing needs of working-class people and is increasingly becoming the reserve of some of the poorest groups in society. This is a process that is often termed *residualisation*.

PRIVATISATION 2: INCREASING RELIANCE ON THE PRIVATE SECTOR

Another form of privatisation is the enhancement of the private sector to meet welfare needs. This can occur either because of inadequate resourcing of the public sector, so that those who can afford it have no choice but to rely upon the private sector, or it can result from positive encouragement for the private sector.

Both of these processes have been in operation in the UK in the last fifteen years in the sphere of health care, for successive Conservative administrations have encouraged the private sector in a number of ways. According to Mohan (1995), the most important factor has been continual exhortation by government to individuals to take out private health insurance. To justify this strategy it has been stressed that public sector health resources are inevitably limited in scope and cannot deal with the infinite variety of health care needs throughout the population. But perhaps equally important have been the substantial waiting lists for many types of surgery that have forced many individuals to utilise the private sector to relieve their pain and anxiety. In recent years various schemes such as fixed-price surgery and loan schemes have been introduced to encourage people to use the private sector for medical care. However, the vast majority of health insurance schemes are provided as an occupational benefit by companies and only 13 per cent of the population in the UK are insured against the costs of medical treatment. This represents a big increase over the figure of about 4 per cent in the 1970s but means that it is still a relatively affluent majority who have access to treatment more rapidly. Furthermore, company funded private health insurance is concentrated amongst the professional and managerial classes, who are disproportionately concentrated in the south of England. As Mohan (1995) notes, this has led to a 'spatial division of welfare' with those in the south-east having access to the upper echelons of a two-tier system.

The distribution of commercial health care very much reflects this distribution of the insured population with a heavy concentration around London. However, in recent years there has been a rapid expansion of the private hospitals in areas that previously had relatively little of this form of health care, notably in the East Anglia and Wessex regions (see Figure 2.1). This pattern of expansion no doubt reflects the growing prosperity of these regions and the out-migration of affluent individuals from the London area.

Another factor that has been crucial in facilitating the growth of private sector hospitals for acute forms of surgery has been the relaxation of controls on private sector practice by NHS consultants. This means that private hospitals have been located close to major NHS facilities to minimise the time spent by consultants travelling between their NHS and private commitments. In addition, there was a relaxation of planning controls on the development of private hospitals.

McLafferty (1989) notes that the last two decades have also seen an increase in the privatisation of hospital services within US cities. Changes in funding regimes have led to increased competition for resources and the widespread closure of smaller hospitals. In response, the non-profit making sector of health care has begun to adopt the strategies of the profit making sector (a process McLafferty refers to as *proprietarisation*). In geographical terms these processes have had highly

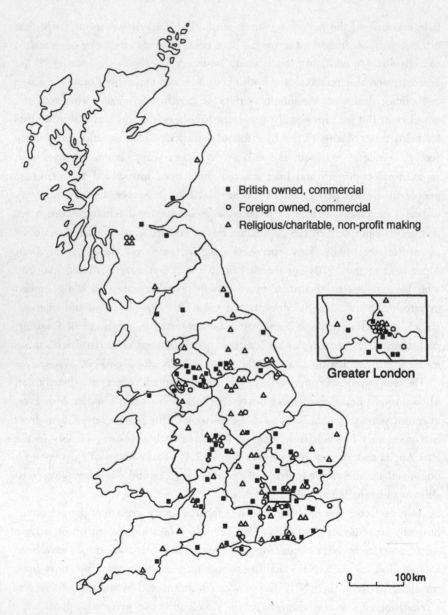

British owned, commercial
Foreign owned, commercial
Religious/charitable, non-profit making

Greater London

0   100 km

Figure 2.1 The distribution of acute sector private hospitals in the UK in 1992

Source: Mohan, 1995, p. 165

uneven outcomes, with closures often concentrated in the most needy low-income areas that are least able to mobilise resources to save health care facilities. In contrast, there has been a proliferation of health services in affluent areas.

Another form of privatisation in the UK has been the private provision of long-term residential accommodation for the elderly. This sector has been encouraged, somewhat inadvertently, by the public sector through the social security budget. An initial lack of restrictions on social security spending meant that entrepreneurs could exploit the run-down of long-stay NHS hospitals for the elderly by providing private accommodation. Thus, many small hotels, guest houses and larger private houses were converted for care of the elderly and this was one of the major sources of small business expansion in the 1980s. Between 1982 and 1985 the number of beds in private nursing homes increased by 143 per cent. This represents a form of *deinstitutionalisation* (see p. 41). Although the provision of this type of accommodation was often supported by the public sector, this was not done through formal tendering and contracting, hence this could not be regarded as a form of *competitive tendering* (see Chapter 3).

The expansion of private nursing homes has been very uneven across the UK (see Phillips and Vincent, 1986 and Phillips et al., 1987). Beds in private nursing homes have been concentrated in areas where there are already substantial proportions of the elderly – often as a result of in-migration of the elderly from other regions – along the south coast of England and especially in urban areas such as Torquay, Worthing and Hastings. This proliferation of private nursing homes has raised the issue of declining standards of care through lack of regulation and supervision – a debate that has been fuelled by some scandals exposed by the media in recent years. The expansion of these homes has also raised the fear that they will encourage further in-migration of the elderly to areas that already have a high proportion of retirees, thus putting excessive strain on local social services. Indeed, some local authorities in the south-west have placed restrictions on the expansion of private nursing homes in certain locations.

In the Canadian context, Laws (1988; 1989) notes how the for-profit sector is becoming an increasingly important component of the welfare system in Ontario. Thus, between 1973 and 1983 commercially operated nursing homes increased the numbers of beds they provided by 30 per cent, whilst over the same period the number of publicly provided beds fell by over 7 per cent. The policy of *deinstitutionalisation* (see p. 41) has also encouraged privatisation. As Ontario shifted away from large institutional forms of care for the mentally ill, the relevant legislation amounted to an invitation to the private sector to participate in the delivery of care.

SELF-PROVISIONING

One of the consequences of rationalisation and asset sales is that an increasing proportion of the population are forced to become self-reliant in meeting their welfare needs. In the case of private sector services this has been termed *self-provisioning* (Urry, 1987). Examples of self-provisioning have included the use of home entertainment rather than live spectacles, microwaveable meals rather than eating out and washing machines rather than use of laundries (Gershuny and Miles, 1983). In the context of the public sector services self-provisioning entails a shift back to provision by families, relatives, friends and others in local communities. Where the services are reallocated back to the responsibility of family labour, this process may also be termed *domestication* (Urry, 1987). As indicated in Chapter 1, familes and members of local neighbourhoods have long been the major source of welfare provision but we have relatively little knowledge of the capacity of different types of community to cope with service cuts.

One trend that can be encapsulated within self-provisioning is the increasing reliance of local authorities upon unpaid volunteers to compensate for staff reductions. These volunteers include those who assist the social services, hospitals and helpers in schools. The development of so-called community care – the closure of large institutions and the provision of smaller, decentralised facilities – has in reality often led to the domestication of services as families (and in particular women) have been forced to take up the burden of responsibility for the disabled, elderly, mentally handicapped and mentally ill (Finch, 1984; 1990). Williams (1994) asserts that the NHS and Community Care Act 1990 has amounted to an assertion of the primacy of family care and self-help.

VOLUNTARISM

Another consequence of reductions in public spending and welfare services provided directly by the state has been a greater reliance upon the voluntary sector. It has been estimated that there are currently between 350,000 and 400,000 individual charities, voluntary bodies and non-profit making organisations in the UK (Hazell and Whybrew, 1993). These organisations are extraordinarily diverse in character but they share in common the fact that they have been 'established voluntarily, that is neither by statute or for profit' (Taylor, 1988). Although the terms 'charity' and 'voluntary sector' are generally regarded as interchangeable, many prefer the latter term to indicate the fact that funding for voluntary organisations derives not only from individuals. Thus, the voluntary sector also obtains funding through donations from foundations, trusts and companies, fees for services and proceeds from investments. However, the increased reliance upon the voluntary sector in recent years has led to increased funding from both central and local government.

Various rationales are used by those on the Right to justify the use of the voluntary sector. For example, it is argued that the voluntary sector can better deal with the diversity of human needs compared with what is perceived to be the standardisation, uniformity and rigidity of public services. It is also argued that efficiency and consumer choice is enhanced by the voluntary sector. Another argument is that the sector encourages independence and self-reliance. There is also a strand of support for the voluntary sector amongst those on the Left who see the sector as closely related to grassroots democracy and therefore able to undermine the dominance of many disadvantaged people by professional groups.

The researchers who have done most to further our knowledge of the geography of the voluntary sector are Gieger and Wolch (Geiger and Wolch, 1986; Wolch and Geiger, 1983; 1986; Wolch, 1989; 1990; see also Wolch and Reiner, 1985). Wolch argues that in the US, in effect, the voluntary sector has become a *shadow state* ' . . . a para state apparatus with collective service responsibilities previously shouldered by the public sector, administered outside traditional democratic politics, but yet controlled in formal and informal ways' (Wolch, 1989, p. 201; see also Warrington, 1995). In Los Angeles County, for example, there are no less than 8,500 voluntary bodies ranging in size from small neighbourhood associations to large trusts and foundations. The scope of their activities is considerable, ranging from social welfare and community services to arts and cultural organisations. Their spending is enormous, amounting to no less than 60 per cent of per capita public municipal expenditures and 43 per cent of average per capita expenditure on public services. In employment terms, the voluntary sector is a fifth of the size of manufacturing and no less than a quarter of the size of retailing.

Although there was no distinct spatial pattern for specific types of voluntary services, in aggregate, it was the middle- to high-income inner-ring jurisdictions that were service rich whilst the lowest income areas, together with some industrial districts, were service poor (Wolch, 1989). The activities of these organisations are extensive, but Wolch questions their ability to fill the gaps left by reductions in public spending. Indeed, some of those organisations dealing with welfare issues are heavily dependent upon government grants and contracts. Whilst some degree of accountability is required for public money, this dependence limits the activities of many of these groups and the extent to which they can pressurise for more radical types of change. In addition, since funding sources are never enough, these organisations have to expend considerable energy in fund raising through entrepreneurship.

In the UK context, the shift away from a state dominated welfare state towards a pluralistic mix of diverse welfare agencies has also placed greater reliance upon the voluntary sector, albeit on a smaller scale than in the US. In particular, the 1990 NHS and Community Care Act has resulted in a shift of power and responsibility away from local authority social services departments towards the voluntary sector. Some have argued that the voluntary sector is not capable of taking on the mainstream role in welfare provision that is intended in the 1990 Act. There is also evidence that the voluntary sector is using voluntary donations from the public to make up for shortfalls in finance for contracts awarded by local authorities (*Guardian*, 8th November, 1995, Society Supplements, p. 9; see Internal Markets in Chapter 3).

Research by Hurd, Mason and Pinch (1995) into donations by directly funded corporate charitable trusts in the UK has revealed geographical inequalities in the funding of the voluntary sector. Donations are concentrated in the south-east, especially in London, and this pattern can be related to the audience the parent company of the charitable organisation is seeking to address. It would appear that those service based companies located in the south-east were seeking to address national audiences whereas manufacturing companies were more likely to give donations close to where their major manufacturing plants were located. The decline of manufacturing in some regions suggests a form of 'inverse care' whereby the depressed regions are least likely to have a corporation capable of meeting their needs. To date, our knowledge of the geography of the voluntary sector supports a statement made by John Stuart Mill in the middle of the nineteenth century 'Charity almost always does too much or too little; it lavishes its bounty in one place and leaves people to starve in another' (John Stuart Mill, 1848, cited in Lloyd, 1993, p. 178).

DEINSTITUTIONALISATION

Although the policy of deinstitutionalisation was introduced in Canada and the US in the 1960s and 1970s, before the current restructuring of the welfare state, the policy has been greatly affected by recent pressures for financial stringency. This has especially been the case in the UK, where the policy was introduced on a wide scale somewhat later than in North America.

Deinstitutionalisation involves the closure of large institutions providing long-term forms of care for needy groups such as the mentally ill, mentally handicapped, elderly ill or disabled, and their replacement by a variety of 'community-based' forms of care. The latter can involve smaller, purpose-built facilities or it can involve care within private households supported by families, relatives and friends and teams of community-based professionals such as nurses, doctors and social workers. Deinstitutionalisation has often been linked with a shift in provision towards care by the voluntary and private sectors as well as within households. The process is therefore linked with the policies of rationalisation, privatisation, voluntarism and self-provisioning noted above.

Like many spheres of social policy, deinstitutionalisation was introduced for many humane and progressive reasons. It is an attempt to overcome the stigma and poor conditions often attached to large institutions such as psychiatric hospitals. It reflects changing attitudes towards mental illness and was made possible by advances in drug therapy and forms of counselling. However, the policy was introduced at a time of severe financial pressures. There is now overwhelming evidence that whilst many large hospitals have been closed very rapidly, the savings made have not been transferred to develop adequate community-based facilities. In the UK, for example, it appears that in a context in which their expenditure is severely curtailed by central government, local social services departments were often reluctant to take over responsibility for the care of ex-NHS psychiatric patients. The result has been that many former patients of mental hospitals have ended up in sub-standard privately rented accommodation and even in prison. An implicit assumption in official utterances about community care is that there is a large band of volunteers who are willing to look after these needy groups. In practice, the burden of caring frequently falls back upon women, illustrating once again the gender bias in the unfavourable impact of welfare reductions. The result has been care 'in' the community but not 'by' the community.

Deinstitutionalisation has been underway for some time and we therefore know more about the geographical consequences of this policy than most other

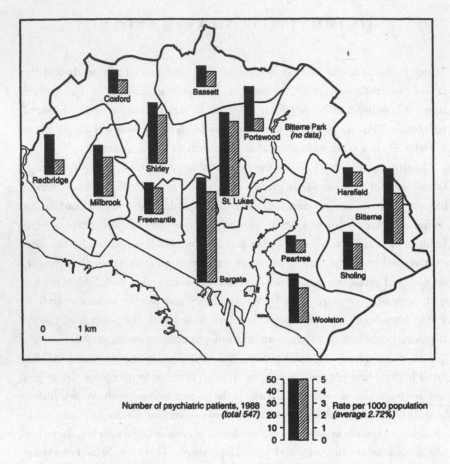

Coxford

Bassett

Portswood

Bitterne Park
(no data)

Redbridge

Shirley

Millbrook

Freemantle

St. Lukes

Harefield

Bitterne

Peartree

Sholing

Bargate

Woolston

0   1 km

Number of psychiatric patients, 1988
(total 547)

50
40
30
20
10
0

5
4
3
2
1
0

Rate per 1000 population
(average 2.72%)

Figure 2.2 The distribution of psychiatric patients in Southampton in 1988

spheres of social policy (e.g. Dear and Wolch, 1987; Laws, 1989; Smith, 1981; Smith and Giggs, 1987; Taylor, 1989; Wolch, 1980; 1981). Given that private residential accommodation tends to be located in inner-city areas, it is these districts that have received the largest concentrations of ex-psychiatric patients. This pattern is well illustrated by the distribution of psychiatric patients in various types of community care in Southampton in 1988 (Figure 2.2), for both the absolute numbers of patients and the rates show a concentration near the city centre. Laws (1989) has also documented the impact of deinstitutionalisation in Hamilton, Ontario. The resulting concentration of community-based facilities in the downtown area led to protests from local residents. These protests, in turn, led the city to pass a by-law imposing a distance separation factor to disperse accommodation for psychiatric patients. However, the effect of this legislation was limited, because there are powerful pressures leading to concentration.

The policy of deinstitutionalisation has been manifest in the most extreme form in the US in California (Dear and Wolch, 1987). The rapid closure of major hospitals has led to the creation of what have been termed 'service dependent ghettos' or the 'asylum without walls' (Wolch, 1981). Former in-patients have been restricted to poorer neighbourhoods whilst zoning restrictions have kept community-based facilities outside of other areas. The process has also been associated with the development of for-profit community-based facilities. Some of these are relatively large with over 100 inhabitants and are beginning to replicate some of the features of the older mental hospitals. In recent years fiscal retrenchment has led to increased problems for many of the mentally ill. Proposition 13 limited tax increases in California, thus reducing the amounts of public money available for community care. Furthermore, the redevelopment and gentrification of some inner areas of Californian cities has restricted the numbers of areas that can be used to develop community-based care. The result has been an increase in homelessness, with people sleeping on the streets in 'cardboard cities'. Indeed, many mentally ill or ex-psychiatric patients are now ending up in prison – a process of *reinstitutionalisation* (Dear and Wolch, 1987).

DEREGULATION

Another important way in which the welfare state has been restructured in recent years is through the process of deregulation. This involves the removal of restrictions on private markets with the broad aim of increasing competition and breaking state monopolies in the provision of welfare services. One approach to deregulation is to make it easier for private capital to enter spheres previously dominated by the public sector. Thus, throughout the 1980s the Thatcher Administrations in Britain encouraged private sector alternatives to public provision in the spheres of health care, education and transport.

One of the most important spheres to be deregulated in recent years has been the labour market. This has taken the form of either abolishing or weakening legislation designed to protect the rights of workers in the areas of the minimum wage, redundancy compensation, holiday and sick pay. The main aim has been to reduce the costs of labour and to increase the flexibility of labour in response to changes in supply and demand. The undermining of employment legislation has been one of the most important changes in the advanced industrialised economies in recent years since it involves one of the main pillars of the welfare state. Not only has deregulation affected the workforce in the private sector but it has also had a big impact upon public sector organisations. It is these changes that we now turn to examine in the next chapter.

CONCLUSION

This chapter has focused upon a wide range of social changes that may be conceptualised as 'chipping away at the edges' of the welfare state. As we have seen, the processes are highly interrelated. Thus, cuts in spending have led to greater self-provisioning and a greater dependence upon the voluntary and charitable sectors. There is also growing evidence that these changes in welfare spending have differential impacts upon social groups based around axes of class, gender and race. In addition, there is growing evidence of the geographically uneven character of many of these changes. The next chapter examines changes in the direct provision of services by the state – what might be termed 'hollowing out' the centre.

FURTHER READING

An overview of welfare restructuring is provided in Pinch (1989). The gender dimensions of welfare cuts are considered in McDowell (1989) and Webster (1985). The best single source on housing sales is Forrest and Murie (1991). A comprehensive review of privatisation initiatives in the NHS is Mohan (1995). The most detailed single source on deinstitutionalisation and its aftermath is Dear and Wolch (1987). The 'shadow state' is analysed in Wolch (1990).

'HOLLOWING OUT' THE CENTRE: CHANGES WITHIN PUBLIC SERVICES

This chapter examines some of the changes to public services that have not been cut or sold off as assets. It is important to remember that a vast range of services still remain within the public sector in most countries. Nevertheless, these public services have been subject to considerable changes in recent years. These changes may be regarded as the 'hollowing-out' of the public sector (see Jessop, 1994). These changes are many and complex and, as will be shown below, are closely interrelated.

INVESTMENT AND TECHNICAL CHANGE

An important facet of the restructuring of the private sector in recent years, especially in the realm of manufacturing, has been investment and technical change (Massey, 1984). This category of restructuring refers to capital investment in new forms of machinery and equipment and is often associated with substantial reductions in employment levels. Relatively little is known about the impact of new technology in the public sector, although a wide range of innovations have been made in recent years. These include the computerisation of health and welfare service records, the introduction of electronic diagnostic equipment in hospitals, cook-chill catering systems, computerised record keeping in libraries and distance learning systems in education (Gershuny and Miles, 1983).

The impact of this new technology upon employment levels in the public sector is difficult to estimate since in many cases new staff are required to establish and maintain the new systems. There is some evidence that the jobs most vulnerable to technological change within the service sector have been clerical jobs, such as when computerised word processing systems have led to staff reductions. Thus, Bradford City Council halved the number of typists it employed through the introduction of a centralised word processing system (Webster, 1985). In a similar vein, larger and more efficient vehicles have been the major reason for cost savings and staff reductions in refuse disposal services (Cubbin et al., 1987). Despite such information, much more research is needed to understand the impact of technological change in the public sector.

INTENSIFICATION

One of the most important changes affecting those who remain within the public sector in recent years has been *intensification* – increases in labour productivity via managerial and organisational changes (Massey, 1984). This type of change has often resulted from redundancies and/or the non-replacement of retiring staff which has served to exert an increased workload upon a reduced work- force. There are numerous examples of this process, one of the most notable being within the NHS where the number of patients treated per employee has risen in recent years. Another example is the higher education system in Britain within which the number of graduates per academic member of staff has increased considerably in recent years (Urry, 1987; Rustin, 1994). The extent of this process of intensification should not be underestimated. For example, in a study of changes experienced by workers in the Southampton city-region in the late 1980s, intensification proved to be one of the most common forms of change and was often reported by public sector workers including teachers, lecturers, doctors, nurses and social workers (Pinch, 1994).

FLEXIBILISATION

In addition to strategies designed to make public sector workers increase their rate of work output, there have also been numerous attempts to increase the flexibility of their working practices. The term 'flexibility' is a highly emotive one meaning very different things to different people and needing careful analysis. Nevertheless, it is possible to distinguish two main types of flexibility strategy – *functional* and *numerical* (IMS, 1986).

FUNCTIONAL FLEXIBILITY

As discussed above, intensification involves people undertaking their existing jobs at a faster rate. However, increases in productivity are also associated with changes in the types of work that staff undertake. Such changes have often been noted in studies of the private sector where they are termed *functional flexibility* – the capacity of firms to deploy the skills of their employees to match the changing tasks required by variations in workload (IMS, 1986). It has been argued that because of increased market volatility and the gathering pace of technological change, shifts in workload are becoming more frequent and intense. This means that functional flexibility is playing an increasingly important role in ensuring the survival of firms. Such flexibility involves firms extending the range of skills of their workers and may involve breaking down the barriers between different occupations.

There has been considerable debate in the UK about the extent to which functional flexibility has been introduced into companies in the private sector. The main finding to emerge from some of the earlier studies was that, despite some substantial changes in new and much publicised manufacturing plants, there was a very limited amount of functional flexibility in manufacturing plants in the UK (Pinch et al., 1991). The concept of flexibility was interpreted as a powerful managerial ideology rather than a set of actual changes. Most of the changes observed were limited in scope rather than part of a grand design. This piecemeal approach was seen as a response to managerial uncertainty and resistance by unions in the plants concerned. However, there is growing evidence from more recent studies of a progressive increase in the scale of functional flexibility (Hakim, 1990).

Relatively little work has been undertaken on the extent of functional flexibility in the public sector. However, there are grounds for believing that the extent of change has been considerable. For example, in the study of workers in Southampton referred to above, respondents were asked if they had experienced any changes in the types of skills they required to do their jobs in recent years (Pinch, 1994). Interestingly, workers in the public sector displayed one of the highest proportions of respondents claiming that they had needed 'more skill' or a 'different skill' in the previous five years. Although some of these changes involved working at a faster rate – strictly speaking a process of intensification – there were also considerable extensions to the scope of jobs. Often these changes involved adapting to new technology.

Changes in the structure of the NHS in recent years have also increased functional flexibility. A good example is the introduction of 'multiple-service' or 'hotel' contracts. These involve the combined provision of catering, cleaning and portering services using functionally flexible workers. Such contracts have been one consequence of subjecting welfare services to the pressures of competitive tendering (see below). Thus, many of these contracts have been introduced by large private sector service firms such as P&0 Total Facilities Management, BET Contract Services and ISS Servisystem. However, in-house teams have also brought about functional flexibility. For example, Doncaster Healthcare NHS Trust agreed to suspend competitive tendering in exchange for about 300 portering, domestic, catering and laundry staff becoming 'generic hotel service workers' (*Health Services Journal*, 1992).

NUMERICAL FLEXIBILITY

Another important element in the IMS model of the flexible firm is *numerical flexibility*. This involves the ability of firms to adjust their labour inputs over time to meet fluctuations in output (IMS, 1986). This type of flexibility can take many forms including the use of overtime, flexi-time, new shift patterns, part-time workers, temporary workers, casual workers or sub-contracting to outside agencies.

One of the most widespread numerical flexibility strategies in the private sector in recent years has been the use of part-time workers (Pinch and Storey, 1992). Whilst their use has decreased in manufacturing, it has increased considerably in services and especially in retailing. Such part-time workers are used to adjust to short-term variations in demand such as throughout the day or the week. It is becoming clear that part-time workers are also being used to adjust to longer-term variations in demand and to replace temporary or casual workers. However, it is important to realise that the use of part-time workers is not simply a flexibility strategy in the pure sense — their use is also part of a cost cutting strategy. It is certainly true that labour flexibility can help to reduce costs but this need not necessarily be the case; similarly, cutting costs can reduce flexibility. In the case of part-time working, however, there is apparently a strong overlap between the strategies of flexibility and cost reduction. The reason for this is that, depending upon the numbers of hours worked and years of service undertaken, part-time workers in the UK are exempt from workers' rights in the field of redundancy payments, sick leave and holiday entitlements. However, it is also important to realise that the use of part-time workers cannot be explained by any simple economic logic. The reason for this is that the overwhelming majority of part-time workers in the UK are women. Because of their discontinuous work histories, lack of sufficient affordable childcare, domestic obligations in the home and lack of alternative types of work, many women either desire, or are forced into, part-time work. Many employers are therefore deliberately formulating jobs as part-time to exploit the demand for this type of work. If the employment regulations and gender-based nature of the workforce were different, it is likely that the need for flexibility would be constructed in different ways – as indeed it is in occupations where men are the majority of workers and where there is a greater emphasis upon shift working.

The question that arises is to what extent has the public sector been using part-time workers as part of a push for increased flexibility in recent years? It is certainly true that the public sector tends to have one of the highest proportions of part-time workers of all industrial sectors. In the case of the Southampton household survey cited earlier (see p. 47), for example, no less than 38 per cent of the public sector workforce was part-time, second only to the 'distribution' sector which had a total of 39 per cent of part-time workers (Pinch and Storey, 1992). Furthermore, these part-time workers are used in a flexible manner to meet demands at certain times of the day or the week. Notable examples are school meals workers, children's crossing wardens, kitchen staff and workers needed to provide continuous social care in the evenings, at nights and weekends. Part-time workers are also used in the public sector on a temporary basis to provide flexibility over longer time periods, as in the case of teachers and librarians during school or university terms. However, it seems doubtful whether this use of part-time workers is part of some recent enhanced flexibility strategy within the public sector. The reason for this is that the use of part-time workers is long established within the public sector. Whereas the recent enthusiasm of the private sector for part-time workers is very much an employer-led strategy, the use of part-time workers has been very much forced upon the public sector in the past because of the difficulty in obtaining full-time workers (the result of the relatively low pay levels in the public sector compared with those in the private).

Despite the above, there is little doubt that the reliance upon a part-time workforce is a useful strategy for many bodies in the public sector that are forced to reduce costs and be flexible. For example, some local authorities have reduced the numbers of hours worked by their part-time school meals staff to evade employment protection regulations. Furthermore, some local authorities, including Kent, Hertfordshire, Somerset and Wirral, have changed the conditions of service for part-time staff by making them redundant and re-employing them to work for shorter hours without the benefit of free lunches, holiday pay or retainer payments in the school holidays (Webster, 1985). Increased use of part-time workers has also been a consequence of competitive tendering and contracting-out (see section below). However, there is some evidence that increased use of part-time workers can under some circumstances lead to increased inflexibility (Cousins, 1987; 1988). Whereas, in the past, low-paid, full-time workers had been prepared to work extra hours to increase their incomes, it appears that in some cases part-time workers with pressing domestic commitments have been reluctant to increase their working hours. It is therefore probably no coincidence that Pulkingham (1992) notes a change in the composition of part-time workers in the NHS: they are now more likely to be below 30 years of age, single and non-householders.

In contrast to part-time working, there are few difficulties in interpreting the shift towards increased use of temporary workers in the public sector. This has been a major strategy amongst welfare institutions for dealing with financial uncertainty and has been especially pronounced in local government, health and education spheres (McGegor and Sproull, 1992).

CONTRACTING-OUT

Another important change in the structure of institutions within the welfare state has been the introduction of *sub-contracting* or *contracting-out* — what is also sometimes termed within the private sector *outsourcing* or *distancing*. Contracting-out involves a situation in which one organisation contracts with another for the provision of a good or service (Ascher, 1987). When applied to the public sector, contracting-out may be regarded as a form of privatisation, since the contract is often awarded to the private sector. This may be distinguished from *commercialisation*, in which provision remains within the public sector, but there is increased funding from private sources (such as through user charges).

It is also important to distinguish contracting-out from *competitive tendering*. In the latter process, various contractors are invited to tender for the provision of a contract which is usually awarded on the basis of some specified criteria such as least cost, highest quality or greatest flexibility. Competitive tendering does not have to involve contracting-out, since an internal department within the parent organisation devising the tenders may bid for, and win, the contract. In these circumstances the contract is often said to be awarded *in-house* – a process that is also termed *contracting-in* or *market testing*. Conversely, contracting-out does not necessarily have to involve competitive tendering. Thus, many companies in the private sector who engage in sub-contracting build up long-term relationships of trust and cooperation which do not involve the regular competitive tendering for contracts. In much of continental Europe there is extensive use of contracting-out of public services but relatively little competitive tendering (Walsh, 1995). For example, France has a long tradition of contracting-out services and placing reliance upon the voluntary sector for social care. Similar patterns exist in the US, Australia and Japan – in all these countries there is substantial use of private and voluntary agencies but often without market testing. However, competitive tendering has become widespread throughout the public sector in the UK in recent years and, at the insistence of central government, has been made compulsory for many services – a policy known as *compulsory competitive tendering* (CCT).

There are many reasons for adopting a policy of competitive tendering. One of the most common is the desire to reduce costs. It is assumed that by awarding contracts to the lowest bidder (subject to certain minimum quality safeguards), this will force the various organisations bidding for the contract to seek the most efficient ways of undertaking the work. There are various strategies used by bidders to reduce their costs and win contracts. In the case of many welfare services these strategies include: reducing the numbers of workers (sometimes through leaving vacancies unfilled); introducing new equipment; and changing working practices (increased numerical functional flexibility and intensification of work). Other reasons for competitive tendering and contracting-out which are more common in the private sector include the absence of in-house expertise and the need for greater flexibility in the face of growing uncertainty over technological developments and market trends.

Given the complexity of public services, it is hardly surprising to find that there are many different types of contract in existence. In some cases a given output of a service may be specified and it is up to the contractor to determine how the work should be undertaken. In other cases, where it is not possible to specify an outcome, the contract may concentrate upon the amount of work to be done. Some contracts involve quite detailed specifications of the nature of the work to be done and others are less specific. Some contracts involve penalty clauses or breach of contracts whereas others are less based upon punishment and attempt to develop cooperation and trust over a longer period.

Contracting-out has long existed in the UK; for example, it was used in the NHS in the 1950s to overcome recruitment problems in areas of labour shortage (Ascher, 1987). However, it was after the election of the 1979 Conservative Administration that contracting-out and competitive tendering began on a wide-spread scale. There have been three main spheres in which contracting-out has flourished. First, in 1983 competitive tendering was introduced for ancillary services in the NHS – catering, laundry and cleaning. Second, in 1988 local authorities were required to subject seven manual services to competitive tendering: refuse collection, street cleaning, building cleaning, catering, vehicle maintenance, grounds maintenance and leisure management. Since 1988 other local authority services have been subject to competitive tendering including computing, housing management, law and finance. A third sphere in which competitive tendering has been developed has been local social services. Local authorities have become purchasers of local social services both from internal departments and also from organisations in the private and voluntary sectors.

If taken to extremes, then contracting-out can involve the public sector comprising a small core of personnel who draw up the contracts which are taken up by the private sector. This frequently happens in the US, and moves in this direction are being considered by some local authorities in the UK including Westminster, Wandsworth and Berkshire. Such a structure is close to the concept of the flexible firm and the 'enabling' state (Ridley, 1988). However, despite all the rhetoric about choice and flexibility in meeting client needs, it is clear that, rather than flexibility, the motivating force behind contracting-out in the public sector has been the desire to reduce costs. As Cochrane (1991) notes, an important part of this strategy has been the desire to impose discipline upon the highly unionised public sector workforce. There have been a number of studies examining the scale of savings derived from contracting-out. Evidence from the US, Japan, Australia, Canada and the UK indicates that between 20 to 30 per cent can be saved through the use of the private sector (see Walsh, 1995 for a review of the evidence). However, it is in the simpler, more repetitive services that depend upon manual labour where the greatest savings can be made. These savings are made in a number of ways. In the case of refuse collection, for example, much of the saving has come from the use of more efficient vehicles (Cubbin et al., 1987). In other services productivity has been increased by changing working practices. However, where labour costs are a high proportion of the total, then savings have emerged from reductions in overall levels of pay. This has come about in a variety of ways including the use of non-unionised labour, increased use of part-time workers and in some cases younger, less experienced workers on short-term contracts.

One of the most fascinating and perplexing characteristics of contracting-out is that it shows wide variations between different areas. Not only are there wide variations between countries but there are also substantial variations for the same service within countries. Moon and Parnell (1986) were amongst the first to draw attention to geographical variations in the privatisation of local authority services which they related to the political complexion of the local authorities. In a later analysis, Painter (1990) examined variations in the first round of competitive tendering in British local government following the 1988 Local Government Act. Not surprisingly, given their opposition to the policy, only a very small minority of Labour-controlled councils contracted-out the requisite services. It appears that such councils are able to prevent contracting-out by a number of strategies (which are considered in greater detail in Chapter 5). In marked contrast, Conservative-controlled councils introduced contracting-out for over half of the services involved. Taking the services as whole, Painter noted four main types of local authority. First, there were the New Right councils who wished to progress towards contracting-out as many services as possible, retaining only a small core of staff to monitor the contracts. These included high profile local authorities such as Wandsworth, Westminster and Bradford. Second, there were the non-committed councils who were prepared to let the tendering process run its course but without major commitment either way. These councils tended to be Conservative-controlled. Third, there was the set of both Conservative and Labour-controlled councils who were keen to retain services in-house. However, they were prepared to make changes to wages and conditions to enforce such a change. Finally, there was the group of Labour-controlled authorities, including Manchester and Sheffield, who were keen to maintain services by in-house teams without cutting wages or conditions.

More recent analysis of subsequent rounds of competitive tendering in local government confirms the existence of a north–south divide in the UK, with private sector companies most successful in winning contracts in London and the south (see Figure 3.1). Patterson and P. Pinch (1995) note a number of factors in addition to political control that can affect the extent to which services are contracted-out. A crucial element is the extent to which private contractors are prepared to bid to work (in some situations there is no competition) or the extent to which their bids are competitive. The scope for realising economies in labour costs is crucial for many private companies and this is much easier to achieve in some services than others. Building-cleaning seems to be the service in which private contractors are able to win the greatest number of contracts through reducing pay and conditions of workers. In some labour markets it is easier to

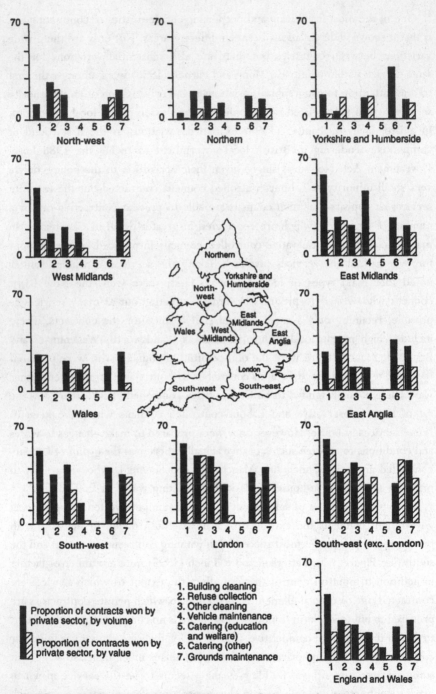

1. Building cleaning
2. Refuse collection
3. Other cleaning
4. Vehicle maintenance
5. Catering (education and welfare)
6. Catering (other)
7. Grounds maintenance

Proportion of contracts won by private sector, by volume

Proportion of contracts won by private sector, by value

Note: Data on catering (education and welfare) are not available on a regional basis

Figure 3.1 The proportion of current CCT contracts won by private sector firms (including management buyouts) in England and Wales in 1993

Source: Patterson and Pinch, 1995

achieve such reductions than others. In contrast to building-cleaning, where expansion requires only hiring additional workers and minibuses to transport them, services such as refuse collection are capital intensive, since there are large thresholds for investment in plant and machinery. This means that larger companies are more likely to bid for such work. The extent to which local authorities are prepared to pass on existing plant and equipment can also affect the success of a bid. There are geographical aspects to the character of work that can also affect the degree of private sector interest. For example, private contractors may be more reluctant to bid for contracts in work dispersed over a wide range of relatively inaccessible sites in rural areas since servicing such a range of facilities will increase costs. In contrast, in compacted urban areas, larger private sector companies can establish large depots from which they can achieve internal economies of scale by serving a number of local authorities. Patterson and P. Pinch (1995) claim that grounds maintenance is an example of a type of contract that is highly attractive to the private sector. In contrast to highly specialist work, such as maintaining gardens in inner-city areas, cutting the grass in large suburban playing fields and parks is a process that can be mechanised and served from a large depot.

Stubbs and Barnett (1992) undertook one of the first geographical studies of contracting-out when they analysed hospital ancillary services in New Zealand. Another study of contracting-out of ancillary services, based on the NHS in the 1980s, also noted geographical variations between health districts (Goodwin and Pinch, 1995). Contracting-out was more likely in health districts in the south-east rather than the north, in urban rather than rural areas, and in suburban areas rather than inner cities. In part, these patterns reflected the bidding patterns of the private contractors. Hence, they were most likely to bid in the south-east where they were already well established serving other private sector clients. In addition, these private companies tended to avoid areas where there were severe logistical problems that would undermine their profitability, such as serving widely dispersed rural areas. However, it was noticeable that some health authorities with similar characteristics, and often adjacent to one another, could display widely differing patterns. Stubbs and Barnett (1992) also found a complex pattern and no single explanation for the geography of contracting-out of ancillary services in New Zealand. Hence, there were only weak correlations between contracting-out and hospital size, hospital location, degree of financial constraint, social complexion of the hospital board, and degree of workforce militancy. Understanding this diversity of contracting-out therefore calls for additional explanatory frameworks, and one major contender for this role – structuration theory – is considered in greater detail in Chapter 5.

INTERNAL MARKETS

One of the more recent changes in welfare services following in the wake of contracting-out has been the introduction of internal markets. These may be regarded as an extension of the market testing approach discussed previously in the section on contracting-out. They are sometimes referred to as *quasi-markets* (Le Grand and Bartlett, 1993) because they mimic the operation of private markets, with various mediating agents such as professionals acting on behalf of the ultimate consumer. They are an increasing component in welfare systems in many countries including Sweden and New Zealand but it is only in the UK that they have been introduced on a compulsory basis (Walsh, 1995).

There are many different types of internal market arrangement. Following Walsh (1995) we can envisage internal markets consisting of three main elements: first, the creation of separate agencies for the purchasing and supply of services; second, the establishment of internal quasi-contracts and trading agreements between these separate agencies; and third, the development of charging and accounting systems. The main justification for internal markets is that they undermine a producer-driven system and provide incentives for efficiency. Purchasers will be forced to make explicit decisions about the most cost-effective way of meeting needs and producers will give incentives to reduce costs and increase efficiency. Of course, the effectiveness of the system depends upon the relative strengths of the purchasers in relation to the producers. In most cases, quasi-markets have been established by public welfare agencies because there are various intermediaries such as professional groups and welfare managers who operate on behalf of the ultimate consumers.

THE NHS AND THE INTERNAL MARKET

The most famous example of an internal market in Britain is that developing within the NHS (Harrison and Wistow, 1992; Hudson, 1992; Mohan, 1995; Salter, 1993). The system involves a division between the purchasers and providers of health care. The former consist of the District Health Authorities (DHAs), who purchase on behalf of general practitioners and budget-holding general practitioners (BHGPs). The funding of the purchasers is based upon a number of need factors including population size. The providers consist of hospitals that are either units directly managed by the DHAs or else are trusts that are independent of the DHAs. The purchasers and providers are linked by various types of contract. The simplest form of contract is of a block character in which the designated hospital agrees to undertake the work directed. More sophisticated contracts involve specifications of the amount and cost of the work undertaken.

The NHS internal market has seen considerable change and experimentation since its introduction and there are considerable variations in arrangements in different parts of the country. In the first year of operation a so-called 'steady state' was recommended and the majority of contracts were between the DHAs and the directly managed hospitals and trusts. In a few years trading patterns have increased in complexity with a growing proportion of contracts being made with a diversity of providers both inside and outside the DHAs' boundaries. In addition, there has been a trend for purchasers to combine into larger units through the mergers of DHAs and the development of purchasing consortia through the collaboration of DHAs and FHSAs (Exworthy, 1993).

The problems of the NHS internal market have received considerable publicity both in academic circles and in the mass media. It is beyond the scope of this book to deal with this huge literature in great depth but amongst the most important problems are the following:

Administrative complexity: the proliferation of contracts needs an increased number of administrators to devise and monitor the system. Contracts have to be specified and monitored, and invoices have to be sent and collected. The increased volume of work has generated many complaints of 'administrative overload'. These transaction costs, as they are termed, are especially high in the case of extra-contractual referrals (i.e. contracts outside the standard block system).

Need formulation: the purchasers have had great difficulty in specifying the amount and type of work that needs to be done. The aim of the internal market is to facilitate explicit decisions about the most cost-effective way of rationing health care to meet health needs, but often there is insufficient information to make such decisions.

Changing power relations: the operation of the system depends upon the relative size of the purchasers and providers in any area. For example, where there is one large purchaser and many suppliers (a situation of monopsony) the former has the power to squeeze efficiency savings. However, producers also have incentives to treat and discharge cases as quickly as possible to make extra income. They also have incentives to take extra cases to increase their income rather than deal with the block contracts.

Contract specification: there have been numerous problems in drawing up contracts. In many cases hospitals have 'overperformed' such that the budgets have been spent before the end of the financial year. In some cases contracts have had to be respecified, in other instances certain types of operation have been cancelled. Purchasers have begun to adopt a more strategic view of health care with the nebulous notion of 'health gain' (i.e. focusing upon the outcomes of health care) in addition to measures of cost effectiveness.

Two tier system: there is growing evidence that budget-holding GPs have been able to jump the queue and obtain faster service for their patients through extra-contractual referrals.

Contract monitoring: given the administrative burdens imposed by the system there is relatively little direct monitoring of the quality of contract.

Supporters of the internal market in the NHS would argue that the system is still in its infancy and that it will take some years for it to produce radical change. There has still been a focus upon acute forms of care rather than the alternative community-based forms of provision.

INTERNAL MARKETS AND SOCIAL CARE

The other main sphere of welfare in which internal markets have been introduced in the UK is social care. Unlike the NHS, this system has one main purchaser, the local authority social services department (LASSD). Within this system 'street level' care managers put together care packages incorporating, amongst many other things, the preferences of their clients. It is assumed that care will be provided by a diversity of agencies including the commercial and voluntary sectors. This is one of the most explicit developments of the concept of welfare pluralism (Johnson, 1987) or the mixed economy of welfare (Pinker, 1992; Langan and Clarke, 1994). This policy has been an attempt to end the system whereby a social security subsidy was available only to those who entered residential care. Thus a key element of the policy is the development of domiciliary alternatives to residential care.

Relatively little is known about how the system is evolving but, once again, early research is showing very different arrangements in different localities

(Charlesworth et al., 1995). In Buckinghamshire, for example, a Conservative-controlled local authority, there has been broad support for the reform of social care and a clear purchaser–provider split has been adopted. A Care Management Team is responsible for the assessment of individual needs and putting together 'packages of care' for clients. The Direct Service Team comprises the local authority's existing provision for the elderly and disabled. A separate Service Development Team is responsible for gathering information on the costs of services and managing contracts. Given that the Direct Service Team is in competition with other providers, then it is put into an environment very different to the previous system. Some local authorities are moving towards completely divesting themselves of all direct provision of social services. However, some local authorities such as Labour-controlled Birmingham have rejected an explicit purchaser–provider split. Here the local social services department has continued with its existing structures of direct provision and there has been an emphasis upon developing quality care strategies as an alternative to market mechanisms (Charlesworth et al., 1995). Indeed, a proposal to put the local authority residential homes into a trust was rejected by the members of the authority. Nevertheless, change was forthcoming in this authority through the greater element of decentralisation of responsibility. Such geographical variations in change again raise intriguing questions about the reasons for these variations.

COMMERCIALISATION AND CORPORATISATION

Another important change affecting agencies of the welfare state is the *commercialisation* of services. In this case the state remains the owner of the service but costs are reduced through the imposition of some form of market mechanism, usually in the form of user charges. If the scale of these charges is such that funding is transferred entirely to the market, and public expenditure is eliminated altogether, then this process is sometimes termed *corporatisation*. Following this approach, publicly owned organisations act as if they were private enterprises – a strategy that has been widely employed in New Zealand. However, in the case of welfare services the user charges are usually less than the full market value. Even when the welfare state was at its zenith, few services were completely free at the source of supply; but in an attempt to increase costs, user charges have increased considerably in recent years.

Advocates of charging argue that if services are free consumers will consume more than they would be willing to pay for and there will be misuse of resources. It is also argued that by putting suppliers in contact with purchasers, even if they are not directly paying but are using vouchers, this will exert a discipline upon suppliers to meet their customers' demands. However, because of the nature of the production process, there are problems in deciding upon the appropriate costs for certain public services. The marginal costs of running some public services, such as public parks, are so low that charging would discourage use (for further discussion of charging issues see Walsh, 1995). Certainly, there is growing evidence from the UK that some groups, such as the disabled, are being deprived of services that used to be provided freely by local authorities because of increased user charges. As in the case of investment and technical change, little is known about geographical variations in charging.

DEVOLUTION

Yet another change sweeping through welfare agencies that is closely related to many other changes noted above is devolution. This involves service providers being established as independent units with control of their own budgets. This approach aims to demolish the large hierarchical structure that used to dominate welfare organisations. Devolution is often associated with contracting-out and the development of internal markets but need not necessarily involve these other forms of change. This approach is in accord with 'New Wave' management theory which stresses the advantages of leaner, flatter ('coat hanger') management structures (see Osbourne and Gaebler, 1992 for an example of this management theory in the public sector and Wood, 1989 for a general critique of this work). As in the case of many other changes noted above, a crucial element of devolution is the creation of performance targets and monitoring of performance levels. Table 3.1 summarises some of the differences between traditional, hierarchical forms of public sector management and the 'New Wave' principles.

Devolution has been widespread in Britain, New Zealand Australia and the Netherlands (Walsh, 1995). One of the best examples of devolution in the UK has been local management of schools (Bradford, 1995). Under the previous system, resources were allocated to schools by education departments of local authorities. However, as with many aspects of the public sector, there was little

Table 3.1 Key elements of traditional and 'New Wave' management theory

Traditional management	'New Wave' management
Centralised, hierarchical structures	Decentralised, 'coat hanger' structure with small core and flat periphery
'Hands on' centralised scrutiny	'Hands off' management, devolution of responsibility for achievement of performance targets down to cost centres, task forces and work teams
Centralised pay bargaining	Decentralised pay bargaining, individual-based, performance-related pay
Large body of staff on permanent contracts	Small core of permanent staff, large periphery of numerically flexible workers
Professional ethos (high standards of technical competence judged by professional peers and based on trust), limited explicit quality studies	Managerial ethos (concern with efficiency and effectiveness through compliance with externally imposed rules and 'quality' surveys)
Client oriented	Customer oriented

Sources: adapted from Flynn (1995) and Stoker (1990)

knowledge of costs of education, and schools could do little to influence the size of their budgets. In recent years schools have been given considerable independence in the ways they can control their funds. The new formula for allocation of funds is based on the number of pupils together with some special needs factors. The aim of introducing this approach was to inject a market element by giving schools a powerful incentive to attract additional numbers of pupils. In addition to attracting more pupils, schools can increase their incomes through commercial activities. This system has made schools improve their efforts to attract pupils through an increased emphasis upon marketing. However, there is usually only a limited choice available to most parents because schools often have a high degree of local monopoly.

Another important form of devolution, this time in the US, is the 'new federalism' instituted under the Reagan Administrations. This is designed to give the states the main responsibility for determining welfare issues. The rationale for this policy is to increase the accountability and legitimacy of government. However, what this amounts to is a classic case of central government attempting to pass difficult issues over to lower tiers of administration (Wolch, 1989).

Jessop (1994) has used the notion of 'hollowing-out' to describe the reduction in the powers of the nation-state through transference to other bodies at other levels. This power is ceded upwards to supranational bodies, across to cross-national alliances of local states, and down to local levels of government. However, the British case seems to provide evidence that contradicts this theory of 'hollowing out'. As Patterson and P. Pinch (1995) argue, within the UK power is being taken away from local government and devolved to non-elected separate agencies, but firm control over these agencies is maintained by the central government. Hence, they argue that 'hollowing out' is a term that might more appropriately be applied to local government in the UK.

DECENTRALISATION

Closely related, but conceptually distinct from devolution is the *decentralisation* of services. Decentralisation involves the fragmentation and geographical dispersal of the service provider units. This can involve financial devolution and the provision of independent budgets but this need not necessarily be the case. Thus, offices can be dispersed throughout a political or administrative jurisdiction, but tight control can be maintained from the centre. The main justification for decentralisation is to make services more accessible and user friendly to their consumers. To facilitate this, administrative units are often based upon neighbourhood units. Like the term 'community', 'neighbourhood' means different things to different people and the decentralised units have varied considerably in size, although 10,000 people is a common number. It is intended that decentralisation should subordinate professional structures to those of the local populace. Hence, the geographical area and the wishes of its inhabitants become the focus for the organisation of work. Decentralisation has become very much in vogue in English local government since the 1980s. Typical examples include the decentralisation of local authority housing offices to save tenants having to travel to central city offices to make complaints. Studies indicate considerable satisfaction with many aspects of decentralisation (Lowndes and Stoker, 1992a; 1992b) but it has been argued that smaller units enable a greater surveillance capability on the part of the local state. Surprisingly, given the geographical basis of this policy, geographers have so far made little contribution to debates on decentralisation of services (see Exworthy, 1993).

CONCLUSION

This chapter has examined some of the many changes taking place within the publicly funded services of the welfare state. Although justified by many criteria such as accountability, responsiveness, diversity and democracy, a constant theme running throughout these reforms has been the desire to increase efficiency and save money. However, in many cases there is a lack of hard, quantitative evidence to justify these reforms and they are often the result of political ideologies. What is also clear is that the impact of these reforms has been very uneven across space. It is often argued that these reforms will help to target resources towards those in greatest need but it is clear from the available evidence that the net outcome of many changes has been to further disadvantage the most marginalised and least well-off in society. Thus, the changes to the welfare state have been part of the broader set of processes leading to increased social polarisation in recent years.

The next three chapters examine possible frameworks for understanding the changing geography of the welfare state. Geographers have in the past tended to rely upon various types of managerialist theory coupled to neo-Marxian frameworks to explain variations in service allocations (see Pinch 1985 for a review of these theories). The role of managers still provides a focus for geographical research but their activities have been interpreted within some new frameworks – notably regulation theory (Chapter 4), structuration theory (Chapter 5) and cultural studies (Chapter 6).

FURTHER READING

An overview of welfare restructuring can be found in Pinch (1989) and Walsh (1995). A good guide to some of the earlier changes is Le Grand and Robinson (1984). Ascher (1987) and Cousins (1987) are crucial for studies of contracting-out. The literature on the NHS is growing rapidly – see Mohan (1995), Le Grand and Robinson (1994). Le Grand and Bartlett (1993) consider quasi-markets.

Part II

EVALUATING CONTEMPORARY SOCIAL THEORIES

4

REGULATION THEORY

One of the most important influences upon human geography in the last decade has been the set of ideas that originated from a group of French Marxian economists known as *regulation theory*. These ideas can be traced back to the early 1970s, but it took some time for them to take root outside of France. However, gradually the concepts were taken up in Germany, Holland and Sweden and, following translation into English, the extent of their influence throughout Anglo-Saxon social science has been truly phenomenal (Dunford, 1990). A close inspection of regulation theory, as will be attempted in this chapter, makes it clear why the approach has become so attractive to many scholars. Yet, arguably, the sheer ubiquity of regulationists' ideas has, at least until recently, led researchers to neglect some of their underlying assumptions, and to view these ideas in a somewhat uncritical light.

This chapter outlines the basics of the regulationist approach, and attempts to convey why it has become so popular. This is followed by a critical review of the approach. Most regulationist work has concentrated upon particular types of manufacturing industry in a limited range of countries. This chapter therefore evaluates the general applicability of regulation theory for understanding changes in welfare structures.

THE ATTRACTIONS OF REGULATION THEORY

There are a number of features of regulation theory that help to account for its remarkable popularity. First, a central concern of scholars working within a regulationist framework has been to understand how nations evolve in different ways. Hence, the relationships *between* states are an extremely important feature of regulation theory; for example, growing international competition is seen as a key factor underpinning much recent change. However, the theory suggests that patterns will take distinctive forms in different countries. This approach provides a sharp contrast with neo-classical economic theory which strives to reveal the universal traits that govern human behaviour in all places at all times. According to regulation theory, the differences in nations are not anomalies from some universal pattern; rather, they are elements that are integral to a whole system of social and economic organisation. This approach has great appeal because when attempts are made to explain why one country differs from another, it is seldom possible to resort to a single explanatory factor; instead it is necessary to invoke a web of complex interrelationships.

A second attractive feature of regulation theory is that it is a very wide ranging approach that holds out the promise of integrating many of the complex and seemingly unrelated changes that are taking place in contemporary societies. Regulation theory thus focuses upon both production and consumption, public and private sectors, what people do at work, and also what they consume in the home (Goodwin et al., 1993). In this respect it is similar to other holistic Marxist approaches, but it seems to offer an escape from some of the limitations of previous Marxist formulations.

One of the criticisms frequently directed at these older Marxist perspectives is their tendency to ignore the role of human actions upon outcomes. This leads to a third attractive feature of regulation theory: a stress – in principle at least – upon the crucial role of social and political struggles in determining the ways in which particular countries solve the problems of capitalism. However, in so doing the regulationists have sought to avoid the opposite extreme, known as *voluntarism*, in which humans have the capacity to achieve virtually any outcome. This latter approach is exemplified in Piore and Sabel's assertion that we live in 'a world that might have turned out very differently from the way it did, and thus a history of abandoned but viable alternatives to what exists'(Piore and Sabel, 1984, p. 38). The search for a 'middle way' between economic determinism and voluntarism has been the Holy Grail for social research in the 1980s, and has given regulationist ideas considerable appeal.

A fourth attractive feature of the regulationist approach is that, like Marx, but unlike some of his adherents in the twentieth century, researchers working within this framework have not been afraid to subject their ideas to empirical scrutiny. Regulationist-inspired books therefore abound with tables, graphs and statistics. Within regulation theory there is a refreshing openness and willingness to incorporate many different ideas from different sources – an outlook that is arguably missing in certain previous strands of Marxian thought.

THE KEY ELEMENTS OF REGULATION THEORY

Regulation theory is not a unified theory but a complex set of interrelated ideas that are used and developed in different ways by different authors. Although essentially Marxian in approach, there is much dispute over the validity of many key Marxist assumptions about capitalism. Some authors distinguish between the 'Grenoble' and 'Parisienne' schools of regulation theory, whilst Jessop (1990) perceives no less than seven schools including West German, Nordic, Dutch and North American variants. But even the same authors use regulationist concepts in seemingly contradictory ways. Despite this complexity, there is no doubt that the individual who did most to develop the ideas in a way that brought them to the attention of the English-speaking world was Aglietta (1979). His approach will therefore form the basis of the ideas outlined below.

Regulationist approaches may be regarded as a reaction to the type of Marxian theory inspired by Althusser (1966) that was previously dominant in France. Althusser's approach attempted to solve one of the basic problems of Marxist theory – how to explain the influence of the state in capitalist society without resorting to a form of *economic determinism* which portrays the state as a straightforward instrument of the ruling class. Althusser's solution, derived from *structuralism*, was to suggest that society consisted of three interrelated levels of organisation: the 'economic', the 'political' and the 'social'. The economic level was conceptualised as the most important, in that it conditions the other levels, but it does not do so in any rigid manner, hence the various levels were regarded as being *relatively autonomous*. The regulationists, amongst many others, objected to this approach because it suggests that structures somehow maintain themselves automatically and independently of the actions of people.

The basic questions asked by the regulationists is 'how does capitalism survive despite all its various conflicts, tensions and antagonisms that it generates?'. These conflicts include those between workers and employers over issues such as wages and working conditions; tensions between different industrial sectors; the problems of organising the monetary and banking systems; ensuring there is sufficient demand for the goods and services that are produced; and determining the appropriate scale of public services and welfare benefits. The answer given by the regulationists is that these conflicts and tensions are resolved, ameliorated and 'regulated' by various social norms, rules and regulations. Taken together, all these factors form a *mode of regulation*. Regulationists argue that capitalist economies are marked by distinctive stages in which different types of regulative mechanisms dominate. For example, it is argued that during much of the nineteenth century there was a *competitive* form of regulation whereas throughout much of the twentieth century a *monopolistic* form of regulation has been dominant.

A basic assumption of the regulationist approach is that capitalism is inherently flawed and full of contradictions. Hence, there will be periods of relative stability when a particular mode of regulation can solve these problems but at other times there will be periods of instability as a new mode of regulation is sought. However, regulationists argue that the form of regulation will vary between different nations, reflecting different cultural traditions, institutional practices, economic structures and social and political struggles. Furthermore, during times of change there will be much experimentation with different approaches. The crucial point is that these solutions do not arise automatically in response to the inherent tendencies of the capitalist system but are the result of human actions, trials and experimentation. Thus, some of the solutions are discovered by accident (Lipietz, 1988). Nevertheless, it is argued that the success of these new regulative mechanisms will be constrained by the logic of the capitalist system. It is this logic, and its relationship with changing modes of regulation, that is the primary concern of regulation theory.

Regulation theory is concerned with relatively long-term changes in society rather than with short-run cyclical variations. Furthermore, it is concerned with capitalism as a general global system and not just with the details of particular countries. The concept used to analyse these broader changes in the character of capitalism over time is termed the *regime of accumulation*. Such a regime consists of a stable set of relationships between production and consumption. This notion is conceptualised at a highly abstract level and is not to be found precisely within any one particular nation-state. Hence, within a general regime of accumulation, there can be different modes of regulation in different counties, although these will often have overlapping characteristics.

Regulationists argue that throughout much of the nineteenth century capitalism was regulated by an *extensive* regime of accumulation. This was a phase in which profits were achieved primarily by increasing the amount of output and expanding the scale of the market, rather than by increasing the rate of productivity of the workers employed. Regulationists argue that throughout most of the twentieth century another type of regime of accumulation has been dominant in capitalist societies. This was an *intensive* regime of accumulation characterised by increases in profits primarily through increasing the efficiency with which inputs are used. This regime of accumulation is termed Fordism. It is further argued that we are currently experiencing such a period of rapid change in which this regime of accumulation is being transformed. In order to understand the regulationists' interpretation of recent changes we need to appreciate the earlier Fordist regime of accumulation.

FORDISM

The concept of Fordism was first developed by Gramsci (1973) but the person who did most to promulgate the idea was Aglietta (1979). Fordism is often used to refer to the factory system introduced by Henry Ford in Detroit at the beginning of the twentieth century to mass produce automobiles. However, as stressed above, within regulation theory this term is used to refer to not only a particular type of industrial organisation and labour process but a whole system of social and economic organisation.

On the labour process side, Henry Ford's factory was an extension of the set of ideas developed by another engineer, Frederick Taylor, known as his 'principles of scientific management'. There were three basic principles: first, that work tasks should be simplified as much as possible; second, that these tasks were to be controlled by managers; and third, that 'time and motion studies' were to be employed to devise the most efficient ways of working. These ideas could be applied to individuals but Ford realised the enormous advantages that would be gained by developing these ideas on a large scale. Ford therefore introduced these ideas on a moving assembly line – previously used for stripping carcasses of meat – and applied them to the manufacture of automobiles. The fragmentation of the assembly process meant that conveyor belts could bring the parts to various stages on the production line. Here the parts could be assembled by relatively unskilled workers who did a limited number of tasks assisted by specialised types of machine tools. There were strong links in this system between the technical division of labour – the tasks to be accomplished – and the social division of labour – the sorts of people who filled these tasks.

Hence, the unskilled work suited the large numbers of unskilled workers in the large industrial cities, many of whom had little knowledge of English.

The extreme specialisation of tasks and machines in this system meant that it was extremely expensive to establish the new production lines. However, once the assembly lines were running, unit costs fell sharply if there were large volumes of output. This created problems because under the extensive regime of accumulation, the market for many goods – such as cars – was limited to the more affluent of the population, while the mass of workers lived in relative poverty. The new Fordist production system brought the capacity for greatly increased output, but if the majority of the population continued to exist at previous standards of living there would be no market for the new goods. Fordism as a system of social organisation was therefore a way of solving this problem by linking production with consumption.

Increasing consumption of the new products was brought about in a number of different ways. Since the mass production system brought about greatly increased productivity and profitability, Ford was able, on the one hand, to cut the price of his products and, on the other hand, to reward his workers with increases in their wage levels. As other employers adopted Fordist production methods, and with the increasing availability of credit, workers were increasingly able to afford the new mass produced items. Fordism has thus been termed the 'commoditization of consumption' (Roobeek, 1987). Although the mass consumption of standardised goods increased considerably after the First World War, this was greatly halted by the Depression of the 1930s. Thus, it was only after the Second World War that Fordism developed most extensively. During this time it became associated with collective bargaining between trade unions and managers. However, a delicate balance had to be kept so that wages were not too high to dent profits and discourage investment, nor too low to dent demand (Jessop, 1990). In contrast to the fierce competition in the earlier extensive regime of accumulation, profits were maintained during the Fordist era by the tendency for price fixing by oligopolistic suppliers.

The regulation of production and consumption also involved the actions of governments. It is at this stage that regulation theory stresses the very different ways in which nations have developed. Thus, some commentators make a distinction between US, European and Japanese styles of Fordism. Much of the Fordist production system was the same in these nations and so some of the greatest difference between them centres around the nature of their welfare states.

The welfare state was least developed in the US. There were numerous reasons for this: the working class was fragmented by racial and ethnic divisions, and there was no unified capitalist class. In addition, the Democratic Party was riven with conflicts reflecting the diverse regional interests in the US. In Europe, in contrast, welfare states developed more extensively. In the inter-war period production and consumption in Germany were regulated by Nazi fascism. The mass production of the Volkswagen beetle (only fully realised after the defeat of fascism) and the development of the Autobahn system had the same significance for German workers as the Model T and the interstate highway for workers in United States (Roobeek, 1987). However, after the Second World War a social democratic form of Fordism developed in Europe. Under this so-called Europeanisation of Fordism, consumption was boosted through collective bargaining which, together with comprehensive social security systems, helped to maintain the incomes of the working class.

THE DEMISE OF FORDISM

This analysis of Fordism is used by regulationists to explain the changes in the fortunes of capitalist economies since the Second World War. During the period from 1950 to about 1973, capitalist economies grew at rates far in excess of previous years in the twentieth century, and this is attributed to the success of Fordism as a mode of regulation. Since the mid-1970s, however, it is argued that Fordism has been in crisis. The extent to which Fordism has collapsed and the extent to which we have moved to a new mode of regulation is a source of some debate. For example, Aglietta (1979) argues that we are witnessing the modification and regeneration of Fordism into 'neo-Fordism'. Others argue that we are witnessing the beginnings of a complete transformation that can be labelled 'post-Fordism'. There is, however, considerable agreement over the type of problems associated with Fordism, even if there is not agreement over their relative importance.

The Fordist system was devised in the United States and regulationists argue that the success of the system was dependent upon the dominance of the world economy by that country. However, gradually the hegemony of the United States was undermined. The nations of Europe and Japan began to rebuild the economies that had been shattered by the Second World War and some developing countries grew at staggering rates. In addition, the monetary system established after the Second World War to bolster the dominance of the US — generally known as Bretton Woods — began to collapse.

The basic problem underlying Fordism as a mode of regulation is declining productivity. Numerous explanations have been forwarded to explain this decline. It is suggested that the market for the standardised products of Fordism became exhausted. In an attempt to overcome the saturation of mass markets in the industrialised nations, companies sought out markets in the less developed world; but this strategy was limited by the increasing debts of the latter. Declining productivity has also been attributed to the failure of companies to invest sufficiently in research and development. Yet another problem was the increasing costs of environmental and safety regulations in the developed countries. There was also the problem of increasing costs of raw materials, in particular, the raised fuel costs brought about by the oil crises of 1974 and 1979 to 1980.

This problem of declining productivity growth was compounded by the fact that, in a context of labour shortage, trade unions assumed a powerful position and were able to negotiate wage increases for their members that were not compensated by increases in productivity. These problems were especially acute in Britain where a multitude of overlapping craft and industrial unions, plus a decentralised and fragmented bargaining system, led to the greatest disparities between wage increases and productivity. One response to these increasing labour costs was the shift of labour intensive work to low-cost locations in the developing countries which served to undermine the productive capacity of the Fordist system in the advanced nations. Furthermore, in the developing countries the system of matching increasing output with increasing levels of real wages was undermined.

There were also particular problems with Fordism as a production system which intensified the above problems. The repetitive and therefore tedious nature of assembly work led to high levels of sick leave, absenteeism, strikes and poor quality of work. There was no in-built system of quality control in the Fordist system and the costs of correcting mistakes after products left the assembly line were considerable. The Fordist system was based upon internal economies of scale which required large outputs of standardised products to recoup the enormous start-up costs of capital investment in specialised machinery. To cope with strikes in factories that provided the parts needed for further assembly in Fordist factories, large supplies of components were stockpiled, which added to the costs of assembly. Furthermore, because of the need for mass production of large quantities of standardised goods, it was difficult to change output quickly in response to changes in consumer preferences.

Further problems in the Fordist system relate to the welfare state. Increasing governmental involvement in services such as health, education and recreation extended their provision beyond the ranks of the poorest in society and into the ranks of the middle classes. The costs of the welfare state became increasingly burdensome, but since a majority benefited, there were strong electoral imperatives that made it difficult to reduce spending.

There have been numerous attempts to solve the control problems associated with the Fordist regime of accumulation. Since the success of many of these innovations is still questionable, many regulationists would argue that we are in a transition phase from one dominant regime of accumulation to another.

NEO- (OR POST-?) -FORDISM

Given the diversity of problems associated with the Fordist mode of regulation it is hardly surprising that many new regulative mechanisms have been identified in the neo- (or post-) Fordist regime of accumulation. In this section I will use Aglietta's term 'neo-Fordism' to denote scepticism about the extent to which we have entered a radically new era of accumulation.

To begin with, it is argued that the saturation of mass markets and the inflexibility of Fordism has been countered by the growth of new, flexible production methods. These new methods, often collectively termed *flexible specialisation*, involve the ability to adjust levels and types of output relatively quickly in response to variations in consumer demand without adversely affecting productivity or profitability. This flexibility of response is achieved in a number of ways. One approach is to use new technologies: flexible manufacturing systems (FMS); computer numerically controlled manufacturing systems (CNC); computer aided design (CAD); and computer aided manufacture (CAM).

Another method consists of flexible labour practices. The rigid job demarcations and fragmentation of tasks associated with Fordism is giving way to functional flexibility (the capacity of workers to vary the tasks that they undertake), as discussed in the context of the public sector in Chapter 3. These extensions in job skills mean that job descriptions are increasingly broad and firms have an increasing capacity to deploy workers between different types of task as and when needed. Another flexible labour practice is termed numerical flexibility – this is the ability of companies to adjust the numbers of workers in response to variations in workload (see Chapter 3). This is achieved by the use of various types of peripheral worker: temporary workers, part-time workers, subcontractors and agency workers. Often linked to the above are new ways of rewarding workers: performance related pay, bonus schemes and the like.

A crucial feature of the Fordist system was control of labour costs. However, in the neo-Fordist era, labour costs are a diminishing proportion of the total cost of production. The size of the workforce needed in most modern factories is relatively small and, in an era of widespread unemployment, employers can therefore be highly selective in who they recruit. Increasingly, an emphasis is being placed, not upon technical skills alone, but on social skills such as reliability. Control of labour costs is therefore less of a priority than it was in the past. But, arguably, much more important than control of labour costs in the post-Fordist era is control over research, marketing and product innovation. To facilitate this, new types of inter-industry alliances are being created. It is argued that all of these changes are providing competitive advantages to smaller firms since they are capable of being more flexible and responsive to market changes than larger, more rigid firms engaged in mass production. Through their use of flexible machines and workers the smaller firms can reduce the costs of customisation through economies of scope.

It has been further argued that these changes have facilitated the emergence of a new geography of production. The geographer who has done most to suggest links between modes of regulation and the geography of production is Scott (1988). He argues that adjustment to the uncertainty and instability of production is enhanced by sub-contracting functions such as research, design, marketing and component supplies. This places increased reliance upon communications between firms. These 'transaction costs' can be reduced by smaller, flexible firms agglomerating together in new forms of industrial district. Hence 'vertical disintegration encourages agglomeration and agglomeration encourages vertical disintegration.' (Scott, 1986, p. 224). This explanation is used to account for the growth of what are termed 'new industrial spaces' such as the dense networks of small companies in the United States (Silicon Valley, Orange County, Route 128); England (the M4 Corridor); France (Grenoble); Italy (Bologna, Emilia, Arezzo); and Germany (Baden-Wurttemberg) (Storper and Scott, 1989).

Table 4.1 is an attempt to summarise this diverse set of ideas concerning Fordism and neo-Fordism.

Table 4.1 Differences between the ideal types of Fordism and neo- (or post-?) Fordism

Fordism	Neo (or Post-?) Fordism
THE LABOUR PROCESS	
unskilled and semi-skilled workers	multi-skilled workers
single tasks	multiple tasks
job specialisation	job demarcation
limited training	extensive on-the-job training
LABOUR RELATIONS	
general or industrial unions	absence of unions, 'company unions' 'no-strike deals'
Taylorism	'human relations management'
centralised national pay bargaining	decentralised local plant-level bargaining
INDUSTRIAL ORGANISATION	
vertically integrated large companies	quasi-vertical integration, i.e. decentralisation,
sub-contracting	strategic alliances, growth of small businesses.
TECHNOLOGY	
machinery dedicated to production of single products	flexible production systems, CAD/CAM robotics, information technology
ORGANISING PRINCIPLES	
mass production of standardised products	small batch production
economies of scale, resource driven	economies of scope, market driven
large buffer stocks of parts produced just-in-case	small stocks delivered 'just-in-time'
quality testing after assembly	quality built into production process
defective parts concealed in stocks	immediate rejection of poor quality components
cost reductions primarily through wage control	competitiveness through innovation
MODES OF CONSUMPTION	
mass production of consumer goods	fragmented niche marketing
uniformity and standardisation	diversity
LOCATIONAL CHARACTERISTICS	
dispersed manufacturing plants in spatial division of labour	geographical clustering of industries in flexible industrial districts
regional functional specialisation	agglomeration
world-wide sourcing of components	components obtained from spatially proximate quasi-integrated firms
growth of large industrial conurbations	growth of 'new industrial spaces' in rural semi-peripheral areas

Table 4.1 Continued

Fordism	Neo (or Post-?) Fordism
ROLE OF THE STATE (see Table 4.2 for variations)	
Keynesian welfare state	The 'workfare state'
demand management of economy	Encouragement of innovation and competition
provision of public services	privatisation, deregulation
protection of the 'social wage'	encouragement of self-reliance
PROBLEMS	
inflation	unemployment
market saturation	labour market dualism
poor-quality products	social polarisation, exclusion and associated
inflexibility	social tensions
alienated workforce	lack of consumer confidence because of insecurity
divergence between rising wages and declining productivity growth	
fiscal crisis of state	

Sources: adapted from Aglietta (1979); Bagguley (1991); Harvey (1989); Roobeek (1987); Stoker (1990)

REGULATION THEORY AND THE WELFARE STATE

The regulationists have had much more to say about the mass production of automobiles than about the functioning of the welfare state, private consumption or social reproduction. Nevertheless, regulationists' ideas have implications for the welfare state and a number of scholars have attempted to develop their ideas.

Aglietta's (1979) book said very little about the public sector but he argued that central to the new regime of accumulation is a new form of labour process in which new technology plays a crucial role. In particular, automation and information technology can replace the fragmented work structure and hierarchical discipline of Fordist production systems with semi-autonomous working groups. These new technologies, he argued, can overcome a central problem of the Fordist regime of accumulation – the divergence between productivity growth in manufacturing and services. Aglietta also argued that new technology would enable state provided services in the sphere of health, education and transport to become available through private markets.

It is important to remember that Aglietta's ideas were written in the 1970s when the implications of new technologies in the microelectronics field were less clear than they are today. Nevertheless, over a decade after they were written, Aglietta's speculations have proved to be influential. The first person in Britain to make links between regulationist ideas and the restructuring of the public sector was Hoggett (1987). Although he wrote a small speculative essay in a relatively obscure publication, Hoggett's essay had a big impact. Whereas Aglietta had argued that the state provides services precisely because they could not be organised on Fordist lines, Hoggett argued that the Keynesian welfare state was inflexibly geared towards the output of a few standardised products with the emphasis upon internal economies of scale. He therefore predicted a shift within the public sector towards smaller organisations with more flexible working arrangements. However, like Aglietta's ideas, this conceptualisation may be criticised for technological determinism and downplaying political processes (Cochrane, 1991; see Hoggett, 1994 for a defence against this charge).

Stoker (1989) also attempted to apply regulationist ideas to an understanding of local government in Britain, but his approach is less wedded to the labour process than that of Hoggett. He argues that the Keynesian welfare state was a key element of the Fordist mode of regulation in that it helped to maintain demand for the products of mass consumption. Stoker also argues that during the Fordist period some of the activities of local government adopted Fordist organisational principles. He admits that production methods developed in the manufacturing industry, such as long production runs and assembly line organisation, are inappropriate for many local services, whether they be in the public or the private sector. Nevertheless, he argues that management principles inevitably reflected dominant Fordist thinking. Three principles in particular are identified: *functionalism* (the division of the organisation around particular tasks and responsibilities); *uniformity* (the provision of services to a common standard); and *hierarchy* (organisation in a number of tiers with vertical paths of responsibility). These principles were associated with the introduction of corporate management principles, systems analysis and programme budgeting.

In a critique of this approach, Painter (1991a; 1991b) notes that the existence of Fordist or pseudo-Fordist production methods in local government is not essential for the existence of the Fordist mode of regulation. Painter argues that whilst some local government activities might be seen to have elements of Fordism, the vast majority do not. Indeed, as discussed above, Aglietta argued that it was precisely because these services were *not* amenable to provision using the organisational principles of mass production that they were allocated by the public sector. Furthermore, as Cochrane (1991) points out, the extent to which principles of corporate management, functionalism, uniformity and hierarchy were introduced into local government in the post-war era in Britain is highly questionable. These principles reflect the formal structures of local government, but in reality there was a lack of uniformity and many competing ideologies. Where Stoker does follow Aglietta, however, is in arguing that neo-Fordist production methods have had a crucial impact upon the welfare state. These methods include information technology, flexible machinery, increased sub-contracting and changes in management practices that facilitate intensified use of labour. On the one hand, these methods can cut costs by increasing productivity, whilst on the other hand, they can open up new areas for private profit, either through the sub-contracting of services to private contractors or through the privatisation of services.

Without question, the writer who has done most to interpret developments in the spheres of politics and social policy in Britain from a regulationist perspective is Jessop (1989; 1990; 1991; 1992; 1993; 1994; 1995). Whilst working within a Marxian perspective, Jessop is very critical of earlier political economy perspectives on the welfare state dating from the 1970s. He argues that these approaches either tend to stress the welfare state as some response to the logic of capital accumulation or else stress the state as the outcome of class conflict. He claims that both of these are unsatisfactory: one tending towards a crude functionalism and the other towards voluntarism. He therefore sees the state as an arena in which various struggles and strategies are worked out. Regulationist ideas are an important part of Jessop's attempts to understand these strategies. He argues that notions of Fordism and post-Fordism can be defined according to four main elements:

1. A *labour process.*
2. An *accumulation regime* – a macro-economic system capable of sustaining growth in both capitalist production and consumption.
3. A *social mode of economic regulation* – a set of social institutions, organisational forms and norms which sustain an accumulation regime.
4. A *mode of societalisation* – a set of institutionalised norms and values that help to integrate the accumulation regime and mode of regulation.

Taking the four factors, he then defines Fordism: first, as a labour process involving long runs of standardised goods; second, as an accumulation regime characterised by a virtuous cycle of mass production and mass consumption underpinned by Keynesian demand management; third, as a mode of regulation embodied in the welfare state; and fourth, as a *mode of societalisation* characterised by a middle mass of wage earners (Jessop, 1991; 1994). Much of Jessop's reasoning echoes previous regulationist interpretations. Thus, he suggests that the growth of Fordist production was stabilised by the Keynesian-inspired management of the economy, whilst the welfare state together with collective bargaining helped to improve living standards and boost demand for consumer goods. However, unlike Aglietta, who referred to 'the intimate relationship between the labour process and the mode of consumption that it shapes' (Aglietta, 1979, p. 160), Jessop argues that the labour process can exist independently of an accumulation regime. In contrast, Jessop stresses that an appropriate mode of social regulation is essential to sustain a given accumulation regime.

Apart from his uncoupling of labour processes from accumulation regimes, Jessop's most distinctive contribution lies in his speculations about an emerging post-Fordist set of arrangements. Once again, this regime is distinguished by the four elements noted above. First, as a labour process, post-Fordism is characterised by flexibility both in the use of machines and the workforce. Second, as a regime of accumulation, it is characterised by continuous innovation to achieve global competitiveness via economies of scope for specialised products. This latter approach is necessary to compensate for the saturation of mass markets and the inability of single nation-states to engage in Keynesian policies of demand management in an era of rapid global financial flows. Third, as a mode of regulation, post-Fordism is characterised by the subordination of social policy to the needs of business. Rather than as a means of improving living standards, wages are therefore seen as a cost to industry, and collective bargaining is replaced by individual and plant level arrangements. (Somewhat uncharacteristically, Jessop is reluctant to speculate on what the fourth element, the mode of societalisation, would look like under a post-Fordist regime.) Jessop thus argues that the Keynesian welfare state (KWS) is being replaced by the Schumpeterian workfare state (SWS). The latter is so called because, like Schumpeter, it is concerned with innovation and, as with workfare, it is concerned to subordinate welfare concerns with the needs of production.

Whilst Jessop's interpretation again finds echoes in previous regulationist work, his approach is distinguished by his emphasis upon the mode of social regulation and the imaginative way in which he connects the various elements. His work is pitched at a high level of abstraction and he deliberately exaggerates the differences between the Fordist and post-Fordist arrangements. However,

Jessop recognises the speculative character of his argument and his approach is open and reflexive. Thus, he acknowledges that the Shumpeterian workfare state can be manifest in various forms. For example, the *neo-corporatist* approach attempts to achieve innovation and flexibility through the delegation of responsibility to various organised interest groups. The *neo-statist* approach, in contrast, involves direct state involvement to achieve a mixed welfare economy. Finally, the *neo-liberal* approach attempts to rely upon market forces via privatisation and deregulation.

Unsurprisingly, Jessop identifies Britain as an example of the neo-liberal strategy. His work has been controversial, however, in the extent to which it has emphasised the functional importance of the policies of Thatcherism in acting as the 'economic midwife' to post-Fordism (Jessop et al., 1988). Whereas the Keynesian welfare state was based upon citizens' rights, universal benefits and rising standards of provision, the neo-liberal post-Fordist state is based upon discretion, means testing and minimalism. Whilst acknowledging that many Thatcherite measures were dictated by issues of political strategy rather than economic rationality, Jessop stresses that their cumulative effect has been to provide long-term structural underpinnings to the neo-liberal strategy. In a similar vein, Moulaert and Swyngedouw argue that there is a fit between neo-conservative governments and the regime of flexible accumulation. They argue that the Keynesian welfare state has been replaced by a two-faced Janus-like state with, on the one hand, an 'entrepreneurial' state characterised by deregulation and privatisation and, on the other hand, a 'soup kitchen' state mopping up the effects of the post-Fordist economy (Moulaert et al., 1988; Moulaert and Swyngedouw, 1989).

THE GEOGRAPHY OF THE NEO-FORDIST WELFARE STATE

Compared with the vast amount of literature that the regulation approach has stimulated on the geography of production, there have been few attempts to extend regulationist ideas to the geography of the welfare state. For example, Storper and Scott (1989) make a few, almost throwaway, statements about the matching of flexible economies and new industrial districts with 'correspondingly' regulatory institutions and ways of life. These are seen to be composed mostly of the sort of business culture that is found in the 'sunbelt' areas of the Unites States.

There is an attempt to dismantle the apparatus of the Keynesian welfare state and to reinforce economic competition, entrepreneurialism, privatisation and self-reliance. This is seen to be manifest in different ways in different environments. For example, there are the individualised modes of consumption to be found within the conservative suburban environments of the US 'sunbelt'; there are the gentrified areas of the world cities, with their associated redevelopments celebrating consumerism; and finally, they cite the alliances of organised labour, business and local government in areas of central and north-eastern Italy (the so-called Third Italy) with their traditions of left-wing governments.

More explicit attention to the links between welfare structures and regulation theory is to be found in the work of Painter (1991a). He contrasts the varying responses of local government trade unions to the compulsory competitive tendering of local services in two British local authorities: Wandsworth and Newcastle-upon-Tyne. Although both of these authorities had experienced considerable deindustrialisation in the 1980s, they were very different environments, the former being a Conservative-controlled council while the latter was dominated by the Labour Party. It is therefore not altogether surprising to find that compulsory competitive tendering resulted in different outcomes in the two areas; in Wandsworth services were contracted-out while in Newcastle-upon-Tyne they have been allocated in-house. However, Painter suggested that these outcomes were not simply the result of differences in political complexion, for the strategies of the local unions had an important influence upon the outcomes. Wandsworth has a long history of antagonistic relations between the council and local government trade unions. This can be traced back to the days when the borough had a Labour council but was intensified by the election of a Conservative council. Newcastle-upon-Tyne, in contrast, developed a consensual approach. However, it is difficult to see the advantage of using the regulationist framework to explain specifically geographical variations in contracting-out in this particular case. This leads on to criticism of the regulation approach.

CRITICISMS OF REGULATION THEORY

Although, as shown above, regulation theory has many attractive features, it has been subject to a barrage of criticism in recent years. This is inevitable when ideas become popular, but the extent of the criticism directed against regulation theory is such that it raises serious doubts about the overall viability of the approach. These criticisms apply to the regulationists' view of the production process as well as to the restructuring of the welfare state. The aim of the next section is to outline these major criticisms and their implications.

OVERGENERALISATION

One of the most common criticisms is that the concepts of Fordism and post-Fordism are too generalised to be useful. Regulationist theory stresses the inter-relationships between elements, but the very broad nature of the concepts leads to inconsistency and imprecision. As Hirst and Zeitlin note (1991), this leads to a number of problems and questions. For example, just what is being regulated in the approach? Is it the general contradictions of capitalism, such as the disjunction between production and consumption, or is it some particular feature of the mode of regulation within a country? This imprecision leads to a further question: what is the relationship between the abstract, high-level concepts such as the regime of accumulation and the lower-order concepts such as the mode of regulation? Is it possible to generalise about these issues from the enormous diversity of circumstances to be found in different countries? In fact, in order to be able to make such connections, regulationists are forced to rely upon a very selective account of events in different nations.

For example, regulation theory tends to ascribe a particular type of labour process as dominant in a particular time period. However, critics point out that during the Fordist period those directly involved in mass production on continuously moving assembly lines were only a small proportion of the total workforce (Meegan, 1988). Even if one also considers those in factories making component parts, one is still typically dealing with only a minority of the working population. Critics of regulation theory thus highlight the great diversity of production processes, even in the so-called heyday of Fordism after the Second World War. These processes include small-scale batch assembly, craft work and continuous production processes (such as in oil refineries and chemicals plants). Supporters of regulationist ideas counter this with the argument that a focus upon a limited number of industrial sectors, such as automobiles, is justified because they were the propulsive sectors with a dominant influence over others in spheres such as industrial relations. Similarly, it is argued that information technology has become one of the leading propulsive sectors in the post-Fordist era. However, this is stated as an act of faith more often than it is demonstrated empirically. As Williams and associates note (1987), even car plants show wide variations from the stereotype associated with Henry Ford's original operation.

LIMITED EVIDENCE

The extent to which contemporary production processes can be described as flexible is also highly questionable. There is a growing body of evidence that indicates that the introduction of functional and numerical flexibility has been highly uneven (Allen, 1988). Williams and his associates (1987) found that in the UK new production technologies such as flexible manufacturing systems (FMS) were being introduced by large firms rather than small firms and were being used to boost existing production techniques rather than innovate new ones. In addition, there appears to be a growing consensus that the impact of FMS has been exaggerated (Meegan, 1988).

The extent to which recent changes in the nature of production are genuinely new has also been questioned. It has been argued that production to order, product differentiation and fragmented labour markets were features of nineteenth-century capitalism in Britain (Clarke, 1988). Hudson (1989) argues that many of the changes taking place in older industrial regions, such as increased part-time working, casualised labour and sub-contracting, herald not a new era of neo-Fordism but a return of older forms of capitalist production. He argues that Fordism never really took hold in many of these older regions and that new working practices are, if anything, *pre*-Fordist, but he is in any case sceptical of typifying changes in regional economies by such broad concepts.

These examples suggest that the very broad concepts employed by the regulationists are inadequate when faced with the complexity of the world. Because of this great diversity to be found in different countries and the lack of fit between theory and evidence, regulationists are forced to introduce many *ad hoc* terms to explain the absence of US-style Fordism in different countries. Hence, the UK becomes 'blocked Fordism', France is 'state Fordism', former West Germany 'flexi-Fordism', Spain and Italy 'delayed Fordism' and Mexico, South Korea and Brazil 'peripheral Fordism' (see Table 4.2; Hirst and Zeitlin, 1991; Tickell and Peck, 1992). Thus, the judgement of Williams and his associates upon Piore and Sabel's similar broad brush approach is also relevant to regulation theory, 'The general problem with meta-history is that it tries to stuff too much into the same bag' (Williams et al., 1987, p. 417). In a similar

Table 4.2 Types of Fordism

'Classic' Fordism

Mass production and consumption supported by welfare state (USA)

Flexi-Fordism

Close cooperation between financial and industrial capital facilitating interfirm cooperation
(Former West Germany)

Blocked or Flawed Fordism

Poor integration of finance and industrial capital, poor industrial relations (UK)

State Fordism

State undertakes a leading role in the creation of mass production through direct control of
industry (France)

Permeable Fordism

Key industries are involved in raw materials processing so limited mass production (Canada,
Australia)

Delayed Fordism

Rapid industrialisation delayed until comparatively recently, aided by cheap labour (Spain)

Peripheral Fordism

Mass production utilising cheap labour with limited welfare states, heavy reliance upon exports
(Mexico, South Korea, Brazil)

Sources: adapted from Hirst and Zeitlin (1991) and Tickell and Peck (1992)

vein, Martin (1990) has argued that Fordism and post-Fordism are ideal type
constructs that 'collapse a wide range of actual forms under a stylised abstraction' (p. 1277). The main problem with this approach is that only evidence that
conforms to the ideal type is considered while other results that do not conform
get subsumed under the general patterns or else are ignored. Thus, counter-
vailing or divergent elements are regarded as anachronistic and residual elements
of an ageing regime of Fordism (Amin and Robins, 1990). For example, Scott
regards the recent experience of the ailing British economy as being as relevant
to the new era of flexible specialisation as the experience of Spain was to the
Fordist age (Scott, 1991).

TREATMENT OF GENDER ISSUES

Regulation theory – especially in the form of Aglietta's early ideas – is also
vulnerable to criticism for its neglect of gender (Bagguley, 1991). In recent

decades women have entered the paid workforce in ever increasing numbers, but in the earlier versions of regulation theory they tend to be regarded as somewhat peripheral to the main economy, with their main orientation being towards the home and the domestic sphere. In the neo-Fordist or post-Fordist eras women are still conceptualised as being in a subordinate position in two ways: first, in service industries that have been infused with Fordist working practices, such as production line techniques in typing pools; and second, as the marginalised workers in the newly polarising workforce in an era of smaller, flexible workplaces.

These assumptions largely reflect the 1970s when Aglietta's ideas were being devised. At this time the *reserve army of labour* theory was a highly influential account of the role of women in the economy. This theory derives from classical Marxism and was given new life by Braverman (1974). The theory suggests that the booms and slumps endemic to capitalism will result in variations in the demand for labour and lead to a pool of workers who are periodically employed and then laid-off as market conditions dictate. Women are regarded as being a major component of this 'reserve army'. However, this theory has come in for much criticism in recent years – not least because it fails to fit the facts. Although women have been more likely than men to be laid-off in some manufacturing industries in recent years, they have entered the workforce in ever increasing numbers, whereas men have experienced increasing rates of unemployment (Walby, 1989). The restricted focus of the regulationists also reflects the priority that they give to certain sectors of the economy and in particular certain spheres of manufacturing that have typically been dominated by men – such as auto-mobile assembly. In this context it is interesting to consider that there are sections of mass production that involve a moving assembly line, such as in the manufacture of domestic electrical appliances, that have not figured significantly in regulationist work. As Meegan (1988) suggests, it is surely no mere coinci-dence that these neglected assembly lines are dominated by women. What these approaches have tended to ignore is the social construction of skill: the ways in which certain jobs become labelled as 'male ' or 'female', not because of their inherent characteristics or skill requirements, but because of whether they have traditionally been done by men or women.

SIMPLIFICATION AND MISINTERPRETATION OF KEY PROCESSES

The regulationist approach might still be acceptable if it captured key processes at work. However, the approach is even more vulnerable in this respect. For example, the ways in which regulationists link production and consumption can also be criticised. In particular, the regulationist interpretation of the Great Depression of the 1930s and the way Fordism is alleged to have solved this has been much disputed. As shown above, the regulationists' view is that rising productivity met limited growth because the middle classes were too small and wages did not match productivity growth. However, others have shown that there were sharp increases in wages after the First World War which led to low rates of profitability and market instability (Dunford, 1990). Clarke (1988) observes that there is little quantitative or qualitative evidence to suggest that 1929 marked the transition from one regime to another. In particular, many of the features of the intensive regime of accumulation, such as the growth of mass consumption, were well-established features of nineteenth-century capitalism. He claims that there is no evidence that the low level of wages was a barrier to capital accumulation at this time. Furthermore, there were substantial increases in productivity in many sectors in the nineteenth century including textiles, metal manufacture, food processing and agriculture.

The regulationists' interpretation of the Fordist period after the Second World War has also been much criticised. Clarke (1988) argues that the immediate post-war period was marked by austerity and not rising real incomes. Recession was thwarted, not by rising wages, but by the Marshall plan, rearmament and the Korean War. It was not until the early 1960s that all the classic features of Fordism were created, but the system involved many problems even during its heyday.

Another difficulty is that, contrary to regulationist thinking, increasing real incomes amongst workers in factories engaged in mass production are not essential for market expansion. Indeed, it has been pointed out that despite the increased wages paid to his workers by Henry Ford, relatively few of them bought their own cars; the bulk of demand came from the farmers of the mid-west (Williams et al., 1987). In any case, it is argued that the extra wages paid to his workers by Ford were necessary to stem the very high levels of absenteeism at his plants because of the boring and repetitive nature of the work. Thus, the development of mass production did not require the majority of manufacturing workers to purchase products themselves. Clarke (1988) argues that rising military expenditure, consumer credit and unproductive capitalist expenditure were at least as important in maintaining accumulation as rising real wages and mass consumption.

The regulationists' explanation for the decline of Fordism can also be criticised. To begin with, the argument that mass production has involved declining productivity increases seems to be inconclusive (Meegan, 1988). The problems of trade union militancy can also be exaggerated. It has been argued that industrial disputes were limited to a few Fordist sectors such as automobiles but were absent from other sectors engaged in mass production, and were also prevalent in sectors that did not engage in mass production (Sayer, 1989).

The extent to which markets have become saturated can also be exaggerated. There is still a flourishing market for many mass produced consumer durables including cars and televisions (Williams et al., 1987). The major threat to many companies in these sectors would seem to come not from the inherent limitations of Fordist mass production but from the inherent limitations of western companies in comparison with their Japanese and Asian rivals. The latter are heavily involved in mass production but in ways that solve many of the problems of western mass production. These solutions include 'just-in-time' delivery of parts by sub-contractors and the use of flexible machines.

Many of these approaches are now being adopted in western capitalist enterprises. However, it is important to remember that these innovations were introduced in the 1940s and 1950 when Fordism is supposed to have been at its height. They were not therefore responses to the inherent limitations of Fordism, but were strategies for dealing with the particular problems that Japanese companies faced at this time. For example, a lack of capital meant that they were forced to keep stocks low and this was a major imperative to develop the just-in-time system. The ability to change the output from a limited number of machines was necessary because of the limited numbers of machines possessed by those companies at that time. Given their paradoxical combination of flexible and rigid factors, the extent to which Japanese companies fit into an era of flexible specialisation is a subject of considerable debate. It could be argued that many of these new forms of inter-industry alliance are far from flexible. In the Fordist era companies could break off dealings with component suppliers relatively quickly, whereas today they are locked into long-term relationships. On the other hand, some Japanese companies are dramatically increasing the rate at which new products can be introduced with given amounts of machinery and workers. Nevertheless, Japanese companies illustrate that economies of scale, both internal and external, are still important in many sectors.

REGULATION THEORY AND INDUSTRIAL DISTRICTS

Lovering (1991) has been prominent amongst those criticising Scott's integration of regulation theory with the growth of new industrial districts. He argues that there is no reason why the decline of internal economies of scale and increasing market uncertainty should lead to increasing horizontal and vertical disintegration. Lovering suggests that whether uncertainty translates into rising or falling internal economies depends upon the context of economic practices and institutions. In focusing upon transaction costs in isolation, he argues that Scott is using a 'simple ahistorical rational calculus' (Lovering, 1991, p. 163) divorced from a particular social context.

In a similar vein, Sayer notes that transactions costs analysis involves a 'profound equivocation' (1989, p. 678) about causality. He argues that whereas the theoretical framework involves a shift from cost patterns towards organisational forms, the empirical applications of the approach involve contexts in which the direction of causality is the opposite and where organisational forms have had a profound impact upon the geography of production. He suggests that organisational forms are the creator as much as the creature of the patterns of costs distributed across production systems.

Lovering and others have also highlighted the limited evidence for the existence of new industrial agglomerations. Within Britain, areas of rapid growth such as the M4 corridor do not contain large numbers of firms engaged in close interactions with each other, as portrayed in the literature on industrial districts. In fact, services have been more important than manufacturing in encouraging the growth of such areas – something about which the regulationists have had relatively little to say. Furthermore, there is growing evidence that Italian, French and North American industrial districts are in reality quite heterogeneous, with different histories, structural dynamics and institutional arrangements (Martin, 1990). Amin and Robin (1990) show that even within Italy there are enormous variations in the structure and evolution of industrial areas. Many industrial districts are in the spheres of textiles, clothing and furniture which are far from the high-technology products embodied in the ideal type of flexible specialisation. Many of the new districts are clusters of small firms providing intermediate inputs to larger firms without access to the technology, finance and expertise that is associated with the industrial districts. In addition, they are often in fierce competition with each other – a far cry from the cooperation that is postulated in the industrial district literature. Some of these industries face threats from larger companies in the traditional industrial heartlands of Italy.

The notion that these areas represent the re-emergence of the craft worker is also highly questionable. Many use female, young and family labour to undertake work that is designated as low skilled and therefore low paid. This is combined with evasion of tax and social security contributions. In sum, the high-technology sectors most strongly associated with flexible specialisation are such a small proportion of total employment that they cannot be regarded as having a hegemonic presence, as suggested in the regulationist literature. Amin and Robin (1990) go so far as to argue that so vague are the current definitions of industrial districts that they could include the types of integrated regional economies that developed around automobiles and other consumer goods in the so-called Fordist era.

Even one of the most attractive features of regulation theory, the focus upon the mode of regulation within particular nation-states, has been criticised. This approach views the world as a series of nation-states arranged in a hierarchy. In so doing it tends to downplay the increasing importance of globalising tendencies. Thus, Tickell and Peck (1992) argue that regulationist approaches subordinate international regulative approaches to those of the nation-state.

REGULATION THEORY AND PUBLIC SERVICES

As shown above, in contrast to the diversity of their claims about production systems, regulationists have had relatively little to say about public services. This means that there are a number of ambiguities and limitations in their statements on these issues. For example, the ways in which privatisation can overcome the problems of Fordism are not altogether clear from the regulationists' accounts. Painter (1991a) claims that Aglietta's views on privatisation involve an incipient technological determinism. Certainly, there is little evidence to suggest that privatisation has seen the reduction in production costs through the application of new techniques. Private companies take over the existing service and tend to provide it in much the same way, albeit with a reduced workforce who have to work harder often for lower pay and poorer conditions.

REGULATION THEORY AND SOCIAL CHANGE

The issue of privatisation raises some of the most important criticisms of regulation theory which relate to its conceptualisation of change. Advocates of regulation theory claim that it avoids, on the one hand, structuralism, in which structures somehow maintain and transform themselves free of human agency and, on the other hand, voluntarism in which humans can achieve anything. However, their apparent solution to these twin sins raises a number of methodological difficulties. In theory, regulationists recognise a diversity of struggles by a plurality of interests in capitalist society. However, Hirst and Zeitlin (1991) argue that in practice the approach usually resorts to an uncritical, class-based explanation. For example, the declining rate of productivity under Fordism is attributed to trade union movements which are seen as embodiments of the broad class struggle. This approach tends to ignore other social divisions based around gender, race or religion. However, more often, when descriptions are made of the transition from Fordism to neo-Fordism, greater weight is given to technological and organisational developments in the production process without appreciation of the ways in which these changes are contested by different social groups. This gives the impression that there is something inevitable about the transition. Thus, Clarke (1988) notes that Aglietta's original exposition gave an important role to the class struggle but, in practice, the regulationists have tended to adopt a structuralist-functionalist model in which there is a phase of structural integration followed by structural disintegration. As Elam notes, 'the history of capitalism remains one where "new" techno-economic forces always do the initial acting and "old" socio-institutional frame-works the eventual reacting' (Elam, 1990, p. 12). This 'implied determinism' (Cochrane, 1991) is apparent in the regulationists' view of privatisation. The approach tends to suggest that privatisation is necessary because of the logic of capitalism, yet in reality privatisation is a highly contested political process. It is therefore impossible to see privatisation in purely economic terms.

Following on from the above, the assertion that there is a 'fix' between the policies of neo-conservative regimes and flexible production systems is also highly questionable. Indeed, there are growing indications of problems in the new industrial spaces such as the failure to transfer technology, to train sufficient workers, or to provide sufficient affordable housing (Tickell and Peck, 1992). As stressed above, there is a strong emphasis upon contingency in many elements of regulationist theory. Thus, Lipietz noted that:

regimes of accumulation and modes of regulation are chance discoveries made in the course of human struggles and if they are for a while successful, it is only because they are able to ensure a certain regularity and a certain permanence in social reproduction.

(Lipietz, 1988, p. 15)

Thus, adherents to regulationist ideas are beginning to recognise some of the problems in new industrial spaces (Scott and Paul, 1990). However, Tickell and Peck argue that this emphasis upon indeterminacy makes it difficult to produce a systematic theory of change. As Cochrane (1991, p. 290) notes, regulationist ideas can become 'the juxtaposition of two typologies with little to say about the dynamics of change'.

CONCLUSION

The sheer weight and diversity of the criticisms directed at regulationist ideas serves as a testimony to their influential character in recent years. But whether one is for or against regulation theory, there is no denying that it has set much of the agenda for discussion. Much of the controversy seems to boil down to two broad issues: first, the relevance of the broader theoretical concepts embodied in the approach; and second, the analysis of change. Even if one is sceptical of both of these aspects of regulation theory, there is a growing recognition of the need to look at the regulative processes at work in capitalist societies. Thus Clark (1992, p. 616) has called for renewed study of 'real regulation' - 'a shift in focus from the design of policy to its implementation through administrative practices and procedures'. Lovering (1991) claims that, despite its shortcomings, the value of the Fordist and post-Fordist literature is that it directs our attention to the links between economic and 'non economic' social processes. Although attempts have been made to speculate about local modes of regulation at the sub-national level (Tickell and Peck, 1992), this type of analysis is so far not well developed. Regulation theory is therefore of most use in analysing the broad context within which national welfare regimes are restructuring. What is needed, therefore, is a form of theory that can analyse variations in the behaviour of smaller political and administrative jurisdictions within nations. One potential contender for this role — structuration theory — is examined in the next chapter.

FURTHER READING

Aglietta (1979) is a key early text on regulation theory but can be heavy going. Much more accessible are the reviews by Dunford (1990), Hirst and Zeitlin (1991) and Jessop (1991). The reader edited by Amin (1994) is also an excellent introduction. Fordism and its problems are reviewed with admirable clarity in Allen (1988) and Meegan (1988). See also Allen (1992) for a clear discussion of post-Fordism. The single most useful source on regulation theory and the welfare state is the set of readings compiled by Burrows and Loader (1994). Useful critiques of regulation theory can be found in Clarke (1988), Sayer (1989), Tickell and Peck (1992) and Williams et al. (1987).

5

STRUCTURATION THEORY

Another important influence upon geography in recent years has been structuration theory. This approach is essentially the work of one prolific author — Anthony Giddens — who has refined and developed his concepts in a series of books (Giddens, 1979; 1981; 1984; 1985). However, structuration theory may be envisaged as just one influential element in a much wider trend in social enquiry — the attempt to develop contextual theory. This is the idea that crucial to understanding human behaviour is knowledge of the settings within which that behaviour takes place. Contextual theory is a very diverse set of ideas but it is essential to any understanding of the changing geography of the welfare state.

THE BASICS OF STRUCTURATION THEORY

Structuration theory was developed by Giddens in an attempt to overcome the major rift that has divided social theory. As we noted in the introduction to Chapter 4, on the one hand, there is an extensive body of literature that concentrates upon the influence of material conditions on social outcomes, thereby playing down the importance of individuals. In its most extreme form this approach is manifest in economic determinism. On the other hand, there is an extensive body of literature that focuses primarily upon individuals and their consciousness as the key to understanding social outcomes. In its extreme form this approach can lead to voluntarism. Giddens notes that whilst most approaches in social science tend, at least in principle, to fall between these two extremes, in practice, they tend to concentrate upon one direction of causality. Thus, many studies of economic change play only lip-service to understanding changes made by people. Conversely, many studies of individuals tend to look at the surrounding environment merely as some sort of static backcloth.

Structuration theory is an attempt to overcome this dualism and instead to replace it with a duality that combines the two elements. However, it has been suggested that structuration is not so much a theory as a set of principles that should underpin social enquiry (Thrift, 1985). The first principle is *recursiveness*. It is argued that social systems do not exist 'out there' independently of people but are made up of the numerous, everyday interactions of individuals (known as *recurrent social practices*). These interactions are affected by various societal norms and the resources at the disposal of the individuals concerned. Each inter-action is affected in some way by what went before and will in turn also influence in some way what comes next. The *system* is therefore something that has to be continually 'performed'. This recursiveness raises the possibility of social change, since the small-scale encounters of everyday life can have long-term outcomes. The second principle is *reflexivity*. This is the simple but important point that people are not automata responding to the broader logic of an economic system but live their lives with a great deal of knowledge about their present circum-stances. However, the approach also recognises that people are frequently moti-vated by forces of which they are unconscious. Thus, many people have practical knowledge but they are unable to articulate how they use this knowledge. The approach also recognises that human behaviour is affected by unintended outcomes of actions as well as by unforeseen circumstances.

Giddens' method of incorporating these everyday practices into long-term change is through a clever distinction between *systems* and *structures*. Systems are the outcome of all these routine and non-routine interactions between people. The structure, on the other hand, is the set of rules, norms and resources that individuals use to negotiate their everyday lives. Giddens argues that there are three elements of systems that affect social interaction. The first is *signification* – the rules that underpin the meaning of language used for communication. The second aspect is *domination* – the power relations dependent upon the allo-cation of resources, property and legal sanction. The third element is *legitimation* – the moral rules used to sanction behaviour. The crucial point is that whilst the system exists in time and space, this is not the case with the structure – it is drawn upon in various ways as and when necessary to create or maintain the system. The structure is thus the medium through which the social system affects the individual, but at the same time it is changed by individual actions since these will affect what happens in terms of feedback loops. Giddens thus refers to the 'duality of structure' – the fact that social structures are both constituted *by* human agency, and yet at the same time are the *medium* of this constitution. This continual interaction between system and structure is called structuration.

Giddens argues that this approach enables one to overcome the dualism inherent in previous approaches. For example, previous attempts to incorporate structural elements into theories that focus upon the subjective interpretations of individuals have tended to incorporate structure merely as a constraint. In these approaches the enabling aspect focuses exclusively upon the intentions of the individuals, and the social structure is therefore not included as an active part of the analysis. In Giddens' approach, however, the individual interactions within the system will impact upon the structure. Giddens argues that his approach also overcomes the problems of functionalist approaches in which individual actions are determined by the actions of the total system without taking account of individual motivations. Giddens argues most forcefully that it is people who have needs and not total systems. Even the most well-established institutions are the result of actions taken by individuals under particular historical circumstances and not because of some abstract need generated by a system. Giddens argues that one can specify the conditions necessary for a particular feature of society to be created and to be maintained but this has to be accomplished by knowledgeable actors and is not some automatic response to wider, systemic needs. This leads to a crucial distinction between structuration theory and functionalist thinking; whereas the latter recognises unintended feedback loops which have an important effect upon the social system, in structuration theory it is only those feedback loops that affect the consciousness of actors that are important (Bryant and Jary, 1991).

Giddens' theories have been attractive to geographers, not only because of their apparent reconciliation of the problems of voluntarism and determinism, but also because of their incorporation of time and space. This emerged most explicitly in Giddens' later work which was, in part, a reaction to criticisms of his earlier formulations. One of the main criticisms of his earlier work was that Giddens places too much emphasis upon everyday encounters and ignores various aspects of systems which govern human contact independently of the actors concerned. In other words, it was argued that there was too much attention upon overt action rather than other types of human agency through inaction. Giddens attempted to escape these criticisms with a larger geographical element developed from the writings of Hagerstrand on time–space geography.

Hagerstrand (1973) noted how a variety of factors serve to constrain human behaviour to localised settings. These constraints include *capability constraints* (the needs of humans for regular sleep and food) and *coupling constraints* (the need to interact with others at certain times to perform activities – such as shopping). Given limitations of time, which affect the amount of space that can be traversed, and the fact that both individuals and activities cannot occupy the same space, human behaviour displays highly repetitive and routine patterns. Although Hagerstrand refers to *authority constraints* which limit the spaces that people can use, Giddens argues that his approach lacks a developed theory of power.

Like Hagerstrand, Giddens' approach recognises that social interaction takes place in distinct geographical and temporal settings, many of which involve highly routine interactions. However, as critics point out, lives are also influenced by others outside of these settings. The crucial question, then, is how do these distinctive settings get bound together into a wider whole. In particular, there is a need to understand how those in particular settings are influenced by those with whom they have no contact. Giddens envisages a hierarchy of local settings – termed *locales* – which become integrated into a wider whole. Until the last few centuries, people had to be present together to transact many aspects of life, but increasingly social processes become stretched over time and space. This stretching is referred to by Giddens as *space–time distanciation*. Various factors have contributed to distanciation – the invention of writing, changes in transportation technology, and electronic information storage and retrieval systems. Interestingly, Giddens argues that city structure has played an important part in space–time distanciation, first through the development of centres with religious and ceremonial significance, and later as centres of commodity production and exchange. However, some geographers have argued that Giddens never fully integrates these various levels of locale. For example, Thrift (1985) argues that Giddens' examples oscillate between micro-level studies and analysis of world systems.

THE STRUCTURATION DEBATE

Structuration theory has generated an enormous debate (e.g. Bryant and Jary, 1991; Craib, 1992; Gregson, 1986; 1987; Held and Thompson, 1989; Storper, 1985; Thrift, 1993). Rather like the controversies centred around the works of Marx, of whom Giddens has been so critical, there has been much debate about what structuration theory *really* means. Thus, some critics have perceived Giddens' approach to be voluntarist, others that it is inherently structuralist. It is impossible to cover these controversies in great detail in this book, but there are a number of important points which can be extracted from the literature that have significant implications for a study of the changing geography of the welfare state.

Given that much of structuration theory is essentially a critique of functionalist thinking in social science, it is not surprising that the nature and value of functionalism has assumed a central place in many of the debates. Some, whilst accepting most of the sins of functionalism, would object to Giddens' complete dismissal of any reference to the needs of social systems. For example, Bryant and Jary (1991) note that Giddens' main criticism of functionalism is its failure to distinguish between the causes and the effects of social actions, thereby neglecting consideration of the actions and motivations of the actors involved. However, they argue that it should be possible to take account of purposive human behaviour and, in addition, to recognise the functions performed by social institutions.

Another major source of debate has centred around the difficulty of applying structuration theory empirically. Although, in principle, structuration theory is an attempt to overcome the dualism in social theory between structure and agency, in practice, it is difficult if not impossible to analyse the simultaneous duality of structures and systems as they become 'instantiated'. Thus, in their empirical analyses, social scientists still tend to look at either the structure or the actions of people. In recognition of these difficulties, Giddens recommends the 'bracketing' of social enquiry to look at either structure or agency. Some have argued that this 'bracketing' perpetuates the very dualism that structuration theory was meant to supersede. It has also been suggested that Giddens' definition of agency is too firmly fixed upon everyday encounters rather than properties of systems that govern conduct independently of the creative abilities of the actors concerned.

Despite the barrage of criticism directed at structuration theory, it is clear from Giddens' writings that he does not envisage the theory as a set of propositions that can be tested in any simple empirical fashion (Giddens, 1991). Instead, he regards structuration theory as an 'approach' – a way of sensitising social researchers about certain questions. Indeed, Giddens does not appear to like the work of those authors who have attempted to apply his work wholesale, but prefers those who adopt certain key tenets of structuration theory – often unwittingly it seems!

Giddens' work and the debates that have surrounded it therefore suggest a number of key principles that should underpin social enquiry. First, and most important of all, there is a need to be aware of the *double-hermeneutic*: not only is there a need to take account of the ways in which the individuals being studied reflect upon and monitor their everyday actions, but there is also a need for social scientists to be aware of their own understandings of these situations. Thus, Giddens (1991, p. 219) notes that 'All social research, in my view, no matter how mathematical or quantitative, presumes ethnography.' Second, and related to the above, is the need to take into account the social settings in which social action occurs. Third, there is the need to avoid the many limitations of functionalism.

STRUCTURATION THEORY AND THE CHANGING WELFARE STATE

Although structuration theory has generated an enormous amount of theoretical debate amongst geographers, there have been relatively few attempts to apply the theory empirically. This no doubt reflects some of the practical difficulties noted in the previous section. Interestingly, those few attempts that have been made to utilise the approach have concentrated almost exclusively upon the changing geography of the welfare state.

DEINSTITUTIONALISATION

One of the first examples to utilise structuration theory in geography was the work of Moos and Dear (1986; see also Dear and Moos, 1986). This work focuses upon the social consequences of deinstitutionalisation in Hamilton, Ontario. Moos and Dear attempt to show the interaction of actors and institutions over time and space. For example, they draw attention to the broad trends that set the context for the policy – the history of the asylum and state control over the mentally ill. Throughout the 1950s improvements in drugs and changing psychiatric theories brought about a radical shift in attitudes towards the mentally ill. However, when combined with pressures for financial stringency, this led to an ill-conceived policy of deinstitutionalisation which led to the creation of ghettos of discharged patients in inner-city areas. The status of the mentally ill had been redefined in a radical way and yet the discharged patients had very little say in this process.

Moos and Dear also show the interactions over a shorter time scale. Residents of inner-city wards in Hamilton began to complain about the concentration of discharged patients in rented accommodation in their neighbourhoods. In response, a local politician proposed that lodging houses for discharged patients should be excluded from residential areas. This proposal aroused considerable controversy and there emerged a complex battle between many actors and institutions over the specification of by-laws to exclude ghettos of mentally-ill persons.

It must be admitted that Dear and Moos' work makes considerable demands on the reader. This is, in part, a reflection of the complexity of the interactions they are trying to convey. However, it is also a reflection of the complexity of the language of structuration theory. Furthermore, as the authors also recognise, the theory provides few guidelines for evaluating the importance of the various elements. Nevertheless, Dear and Moos do manage to convey some of the spirit of structuration theory through their study of interactions between individuals and institutions in their distinctive spatial and temporal settings.

SCHOOL CLOSURES

Another example of the application of structuration theory to the study of the changing welfare state is Phipps' analysis of school closures in Saskatoon and Windsor, Canada (Phipps, 1993). Phipps describes a paradoxical situation in which communities threatened with the closure of their local schools in Saskatoon acquiesced to these decisions with relatively little protest. As in the case of deinstitutionalisation, this was in large part a reflection of asymmetry in power relationships. Key levers of power lay in the school boards and their trustees through the rules and resources at their disposal. The school boards exploited their legal powers to facilitate school closures and to constrain the activities of the community representatives. For example, the early approach to school closure had utilised joint educator-parent committees, but these were relatively inefficient in that they enabled articulate and educated community representatives to mobilise against closure. This approach was therefore replaced by public meetings over which the school boards could exercise more control.

Phipps attempts to convey how the system is 'instantiated' through application of the structure – the rules and resources at the disposal of various actors. Given that all the closures went ahead in the settings, what Phipps' study does not convey, however, is the capacity of human agents to change their lot even when the rules and resources are not in their favour.

CONTRACTING-OUT OF NHS ANCILLARY SERVICES

A final example of how the broad framework of structuration theory can be used to illuminate the changing geography of the welfare state is to be found in the work of Goodwin and Pinch (1995). They attempt to explain variations in the geography of contracting-out of NHS ancillary services, as discussed in Chapter 3. As noted previously, whilst there are broad trends in the contracting-out of ancillary services, such as a rural–urban and north–south split, there are wide variations between districts with apparently similar characteristics (see Figure 5.1). In attempting to explain such variations, Goodwin and Pinch utilise the language of structuration theory together with the approach of Pettigrew and associates (Pettigrew et al., 1988; 1990; 1992). The latter shows remarkable similarities to Giddens' approach, even though it was apparently devised independently. Pettigrew and associates note that previous studies of change in the NHS have concentrated upon national barriers to change and have tended to ignore local factors that either enhance or inhibit change. To analyse such local factors they argue that studies should incorporate a number of characteristics. First, studies should be *processual* (i.e. focusing upon action as well as structure); second, studies should be *pluralist* (analysing the competing versions of reality envisaged by different actors engaged in the process of change); and third, the approach should be *historical* (taking into account the development of change as well as the constraints within which the policy makers operate). Crucial to the Pettigrew framework is a distinction between the *context* and the *content* of change. The content consists of the essential characteristics of the change or, in other words, the nature of the policy. The context consists of two parts: first, an outer component, which incorporates the national political and economic context, and second, an inner component, which incorporates the local managerial and labour market cultures. Another element in the Pettigrew framework is the *process* of change which refers to the actions of the individuals

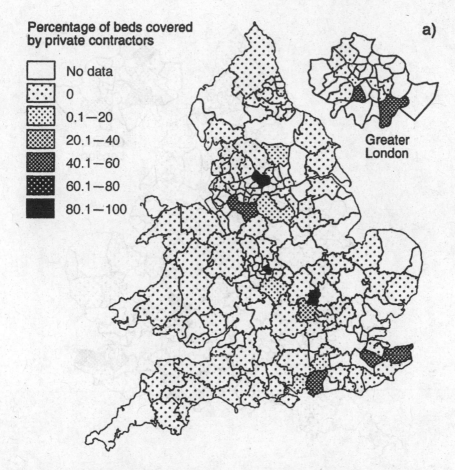

Percentage of beds covered by private contractors

- ☐ No data
- ⬚ 0
- ▧ 0.1—20
- ▨ 20.1—40
- ▩ 40.1—60
- ▦ 60.1—80
- ■ 80.1—100

a)

Greater London

Figure 5.1 The geography of contracted-out domestic cleaning services, 1985–1991: *(a)* Percentage of beds covered by private contractors in 1985

concerned, their perceptions and values and how these are played out in a given context. The crucial point is that the context may be a critical shaper of the process of change whilst at the same time the content of the process may impact upon the local context.

This framework has some striking similarities with Giddens' structuration theory. First, the Pettigrew approach also incorporates *embeddedness* – the need to study change through the interconnections at various levels of analysis. Second, it places in the forefront of the analysis the temporal interconnection of processes. Third, a key concern is to analyse how context is a product of action, whist action is also a product of context. Hence, the idea of process is broadly analogous to Giddens' idea of recurrent social practices, whilst the inner

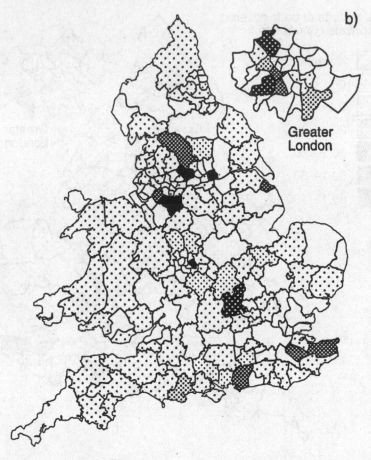

Figure 5.1(b) Percentage of beds covered by private contractors in 1987

and outer contexts help to form the structure upon which the actors draw, either directly or indirectly. Fourth, a central assumption of the approach is that change is not a simple linear phenomenon.

The above might seem abstract but can readily be illustrated with some case studies. Goodwin and Pinch (1995) attempted to analyse the interconnections between context, content and process of change by selecting two district health authorities with very differing histories of contracting-out. The two district authorities (termed A and B to preserve confidentiality) were areas of urban decline in northern England characterised by unemployment, overcrowding and high proportions of single-parent families. Both districts had strong labour movements, with vigorous trade unions. However, despite these similarities, these

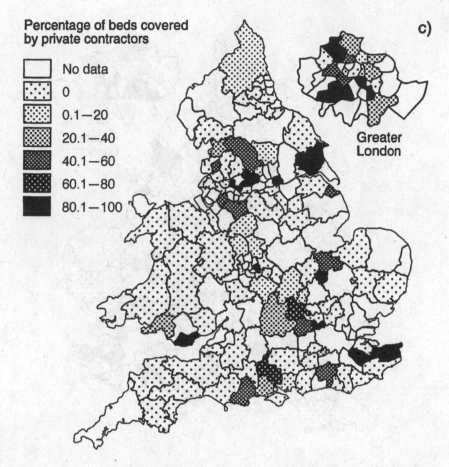

Percentage of beds covered
by private contractors

☐ No data
0
0.1—20
20.1—40
40.1—60
60.1—80
80.1—100

c)

Greater
London

Figure 5.1(c) Percentage of beds covered by private contractors in 1989

districts had very different experiences of contracting-out. In case A, apart from one small contract for domestic cleaning in 1985, all provision was in-house until 1991. In case B, by contrast, all the four main ancillary services of domestic cleaning, catering, laundry and portering were contracted-out between 1983 and 1992. Detailed analysis of these two districts revealed that there was no one single factor responsible for the different outcomes but rather that the initial external pressures for change were mediated by differing managerial cultures in each district.

In case A there was hostility to contracting-out not only from the workforce but also from the management. Thus, it was only after threats of legal action and financial pressures that procedures for competitive tendering were introduced.

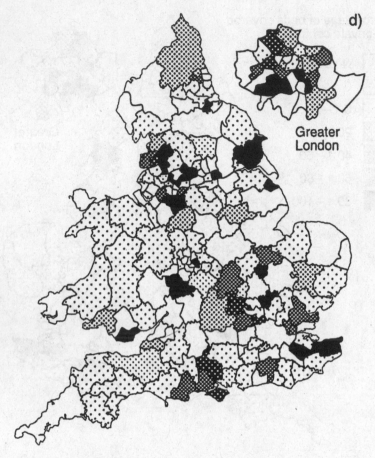

Figure 5.1(d) Percentage of beds covered by private contractors in 1991

The first small contract for the cleaning of twelve health centres and five clinics produced some low, loss-leading bids from the private sector. But such was the anti-privatisation culture in the district that the members voted to keep the service in-house, even though this bid was more expensive. However, the region made it clear that this decision was unacceptable and the district chair used his special powers to award the contract to the cheapest of the private sector bids.

The imposition of this policy from above aroused considerable hostility amongst workers, management and district members. Thus, following the decision to award the contract to a private company, 70 NHS cleaning staff went on strike and this was supported by the TUC- (Trades Union Council) nominated members of the health authority and local labour councillors. The strike also gathered support amongst other ancillary staff who feared for their own jobs, and so catering and portering staff came out in sympathy. After eight weeks of industrial action by 1500 staff a compromise was achieved whereby competitive tendering was suspended provided the unions undertook no further action but accepted the initial private sector tender. It was agreed that those workers made redundant by the contracting-out would be re-employed elsewhere in the authority.

This agreement was of dubious legality but the delay was important because it gave the in-house teams time to develop more attractive bids when the next round of tendering began. There were also important changes in the context within which competitive tendering took place. In the mid-1980s, private sector contractors were overwhelmed with possible contracts and did not have the resources to bid for all of them; hence, they became more selective. Given that there were opportunity costs in bidding for contracts that were subsequently awarded to in-house teams, the private contractors tended to avoid districts where they knew hostility to their presence was likely to be encountered. Thus, when in 1987 all the contracts were put up for tender in case A, in-house teams won all the contracts.

In marked contrast to case A, case B was one of the leading innovators of competitive tendering in the 1980s. Given that the district had a similar socio-economic background to case A, this raises the issue of how this policy was possible. There certainly was hostility to competitive tendering in case B, and there were a number of one-day walkouts when the policy of tendering-out was introduced. However, this opposition was never mobilised into prolonged and concerted opposition. One of the key reasons for the difference in the two case studies seems to be the attitude of management. Unlike managers in case A, those in case B were highly motivated to introduce competitive tendering. The management realised that they would encounter serious industrial relations problems if they implemented the policy without the consent of the staff, so they made considerable efforts to convince the existing workforce of the inevitability and desirability of this policy. Hence, they sought to undermine the key NUPE (National Union of Public Employees) trade union shop steward by circumventing the usual channels of negotiation and by talking directly with staff, through both informal meetings and formal meetings without trade union representation. The management convinced the staff that acquiescing to competitive tendering was in their own best interests and was the best way to preserve their existing jobs. They praised the staff for their efforts in the past and informed them that with new approaches savings could be made for the benefit of the hospitals. This approach proved to be successful and the staff regarded this as more likely to preserve their jobs than the non-negotiation stance of the NUPE representatives. Consequently, the NUPE members voted to oust the existing shop steward and to replace him with someone prepared to negotiate with the management. The majority of the initial tenders were awarded to in-house teams after this change. Significantly, the one group of workers who refused to make concessions to management – the porters – found their in-house bid uncompetitive compared to the private sector and their jobs were awarded to a private tender.

Once again, case B illustrates the temporal interconnectedness of change. Many of the private sector companies were inexperienced in bidding for NHS work in the early 1980s and the work was subsequently awarded to in-house teams. The trade unions were also able to exploit the inexperience of the private sector in drawing attention to deficiencies in the quality of services that were contracted-out. However, over the years the private sector also learned lessons, and in the late 1980s and early 1990s an increasing proportion of contracts were awarded to the private sector in case B. Following the award of trust status in case B, there is evidence that the management became overconfident and aggressive in their implementation of contracting-out. Following the revelation of financial mismanagement, the Chief Executive resigned and the new management team adopted a more conciliatory approach. Once again, there was a concern with discussing change with the workers concerned rather than imposing change from above. Thus, the unions have been able to play an active role in the specification of tenders. Significantly, under the new management regime one of the domestic cleaning contracts and the portering contract were brought back in-house.

THE INFLUENCE OF CONTEXT IN CONTRACTING-OUT

These two case studies illustrate well the value of a contextually sensitive approach in understanding organisational change. In Pettigrew's terminology, one can see the interplay between the content, context and the process of change. For example, in case B changes in the nature of the outer context brought into question the legitimacy of the policy of contracting-out. The outer context involved the EC (European Community) Acquired Rights Directive of

1977 and the corresponding UK Transfer of Undertakings (Protection of Employment) Regulations 1981, commonly known as TUPE. This law states that if a commercial undertaking is transferred, a new employer must safeguard the pay and conditions of the existing staff or face claims of unfair dismissal (Cohen, 1993). The 1981 Employment Act had exempted non-commercial activities from these regulations, thus making possible competitive tendering of NHS support services. However, in 1992 a case at the European Court of Justice resulted in a change to UK law such that TUPE now had to apply to public sector activities. In case B the unions used this new ruling to threaten management with legal action if new arrangements for portering and cleaning threatened the terms and conditions of the employees. The uncertainty over whether the new contract would infringe TUPE, together with knowledge that the legality of the new arrangements would in any case be tested, forced the management in case B to suspend competitive tendering of portering and cleaning in favour of a directly run service.

The local context also had an important influence upon the tendering process in the two districts. Thus, in case A the anti-privatisation culture resulted in the mobilisation of workers and trade unions into sustained industrial action. The local environment in which the new management team took over the trust in case B also influenced development. Faced with considerable bad publicity for previous mismanagement, the team had to adopt a policy in which cooperation with trade unions was essential.

It is also possible to discern in these case studies the way in which the process of change has had an influence upon the contexts in which contracting-out operated. For example, in case A, the long industrial dispute resulted in the creation of an environment that was unattractive to private sector firms. In case B the trade union mobilisation was crucial in fostering the bad press the Trust received at both national and local levels.

These case studies therefore illustrate the ways in which the context can play an important part in shaping the actions of individuals, whilst at the same time these actions can change the nature of the local context. There are some very broad but clear parallels here between the duality of structure formalised in Giddens' structuration theory. Thus, the outcome of one set of interactions serves to form the structural (or contextual) conditions for the next set of interactions. The crucial point to note here is that the context is not simply some static backcloth or outer shell. Instead, it is composed of what Giddens terms 'rules and resources recursively involved in institutions' (Giddens, 1984, p. 24). In the context of competitive tendering this involves legislation, financial resources and incentives, and people's attitudes and knowledge of their predicament.

The studies also show that a crucial element amongst the inner context was the organisational cultures. However, these are seldom monolithic but are typically diverse and highly contested by different factions. Change was not therefore imposed upon these organisations in a single, economically rational or logical manner. There was an ideological struggle for hearts and minds within the two organisations.

The case studies also illustrate the capacity for human agency in the contracting-out process. It is certainly true that some contexts are more likely to facilitate change than others. However, the diversity of outcomes in apparently similar situations reveals the importance of key actors (including managers, trade union officials and workers) upon outcomes. Thus, a receptive context for contracting-out may be reversed by the removal of a key individual or the implementation of an ill-conceived plan. It is therefore not possible to 'read off' in any simple or direct manner the nature of change from the local context or the history of the locale.

These case studies also illustrate Giddens' view that those on the receiving end of a policy imposed from above are not entirely powerless. Thus, the trade union officials and workers in both case studies displayed considerable awareness of their situation and learned a number of lessons as the process of compulsory competitive tendering (CCT) developed. Various strategies of resistance were devised, and although working practices changed, contracting-out was not inevitable.

CONCLUSION

Structuration theory has fallen from favour somewhat in geography in recent years and yet it continues to wield a considerable indirect influence upon research through some of the basic principles that it expounds. There is little doubt that it provides a powerful set of principles that should underline all social enquiry. However, far more important in geography at present is the development of cultural studies. The trend will be considered in the next chapter, where it will become clear that cultural studies is attempting to grapple with some of the basic issues that are central to both regulation and structuration theory.

FURTHER READING

It has to be admitted that reading much of the literature on structuration by both Giddens and his critics is hard work. However, in addition to Giddens' own work cited above, the reader by Bryant and Jary (1991) – including a reply to his critics by Giddens (1991) – is a fairly accessible introduction. See also Cohen (1989) and Craib (1992).

6

THE CULTURAL TURN

Yet another important influence upon human geography in recent years has been the influence of cultural studies. This trend has become widely known as the 'cultural turn' and is also sometimes referred to as the 'linguistic turn'. Although this trend has so far had relatively little impact upon study of the changing geography of the welfare state, it is almost certain to have a big impact in the future. This chapter therefore examines the basic elements of the cultural turn and considers how they can be utilised in studying the geographies of welfare services.

WHAT IS THE CULTURAL TURN?

Cultural studies is sometimes portrayed as some radical post-modern departure from the modernist theories of the past such as the Marxian-inspired theories of political economy (including regulation theory). It is true that in certain forms the cultural turn amounts to a rejection of much previous work. However, it is important to realise the strong links between much that goes under the label of cultural studies and previous work. Inevitably, cultural studies evolved from this previous work, even if it was often a reaction to what were seen as the limitations of these earlier approaches. A brief discussion of the evolution of cultural studies is therefore useful at this point.

As was suggested in Chapter 4 on regulation theory, much recent work has been an attempt to avoid the pitfalls of Althusser's structuralist approach (Althusser, 1966). Althusser was attempting to move away from the classic, Marxist, materialist approach which argued that it was not ideas but economic forces that were the main driving force behind social change. Althusser disputed this orthodox view that cultural artefacts were simply determined by the economic base. He argued that ideology was determined by the economic base but only in the 'last instance'. In other words, the ideological superstructure was an active element in society that could have powerful effects that were 'relatively autonomous' from the economic base. The main implication of this approach was that political order was secured as much by consent as by coercion. In this respect, Althusser was very much influenced by the writings of Gramsci (1973) who attempted to analyse the subtle mechanisms of state control in a capitalist society. This led to a focus upon the role of various 'state apparatuses' involved in mobilising consent, such as the Church, family and education system.

Althusser's approach was also influenced by the structuralist approach of Lévi-Strauss (1969). Structuralism disputes the view that individuals are logical, independent beings who are in control of society. This latter view reflects a philosophical approach that stretches back to Descartes who argued that people are rational and coherent. Structuralism, in contrast, drawing upon Freudian theories, argued that identity is derived from social conventions and deep unconscious motivations that lie hidden. People may think of themselves as the source of all meaning and action but structuralists argue that behaviour is determined by deep, hidden forces. For Lévi-Strauss, the pioneering anthropologist, these underlying forces were universal taboos, but for Althusser these forces were the economic imperatives of the capitalist system. From this perspective, individuals are seen as being caught up in a web of structural relationships of which they may be only dimly aware. It follows, therefore, that social order is maintained, not so much by strict indoctrination or force, but by subtle processes of socialisation. People may therefore take the view that they are rational and independent when they are not. According to a structuralist perspective, the structure places people in particular places to perform certain roles, making them subjects of a broader system.

Many have objected to the structuralist view of society. For example, it tends to downplay the ability of individuals to affect their future. This capacity of people to be active and creative elements in their future is sometimes termed *human agency*. Indeed, as we have seen in Chapters 4 and 5, it is the capacity for human agency that both regulation and structuration theory have been attempting to theorise. In contrast, earlier studies in the structuralist mould tended to represent a rather static view of structure as some fixed, underlying element. Furthermore, many disputed Althusser's view of the relationship between the ideological base and the superstructure. In theory, Althusser's approach avoided the pitfall of reductionism, but in practice, since the 'last instance' never comes, it tended to foster a relative view of culture. Despite the many limitations of structuralism, this approach led to issues of culture being taken much more seriously. Culture was seen as an active element in the social system and no longer a superficial manifestation of economic relations.

Central to the formation of culture is the use of language. Recognition of this has led to what is known as the 'linguistic turn' and another structuralist influence, this time derived from the work of the Swiss linguist Ferdinand de Saussure. His analysis is central to an understanding of the cultural turn. He argued that the meaning of words derives not from the objects to which the words refer but to the differences between related concepts. Furthermore, he asserted that these differences are governed by the rules of grammar. This view might seem contrary to common sense but can easily be grasped; for example, a red traffic light has no meaning in itself – it is the crucial difference between red and green that creates the meaning (Giddens, 1989). The crucial point is that meanings are created internally within the language. A language is, therefore, a set of social conventions. It follows that there is no straight relationship between the world and the words used to depict the world. We have to understand the culture of the society to understand the use of the words within the language.

Culture is often thought of as sets of shared meanings. In general terms culture consists of three elements: first, there are the *values* or abstract ideals that are held by the group; second, there are *norms* – the rules and principles that people are expected to observe; and third, there are the *material goods* that people use (Giddens, 1989). The last element is important in a geographical context. Whilst cultural meanings are often expressed through speech and writing, they can also be expressed in material objects, including buildings and landscapes. All of these elements generate cultural meanings and signify shared values. In the language of cultural studies, the *signifier* is the word, or sign, or material artefact that points to the *signified* (the cultural meaning or message).

The study of the meanings behind signs is known as *semiology*. Structuralists hoped that by exposing the conventions that underpin language, the material base of society could be revealed. They wanted to show how people inherited sets of conventions that were embodied in language. Meanings were not therefore universal elements for all times but 'social constructs' capable of change. By actively participating in the construction of these meanings it was argued that people could bring about revolutionary change. The structuralist approach therefore argues that the authors of written material should be regarded no longer as independent agents but as producers of meanings within a wider set of structural interrelationships. However, early versions of this approach tended to portray these interrelationships in a static, mechanistic way such that the author was represented simply as a prisoner of language. Later work portrayed the author as producing sets of meanings (termed *discourses*) which interacted with other sets of meanings or discourses.

It is at this juncture that cultural studies becomes embroiled with the growth of post-modernism. Post-modernism is, amongst many other things, a reaction to the universal, monolithic theories of the past which purport to explain everything in society (such as Marxism). Under the influence of such theories, structuralists had tended to relate cultural meanings and signs to some under-lying, unchangeable scheme. However, the idea that culturally derived sign systems could be explained by some universal approach (known as a *metalanguage* or *totalising narrative*) became questioned. One of the early proponents of a change was Barthes (1973). Unlike the early structuralists, he did not regard himself as presenting privileged insights into universal objective truths. He argued that social systems and sets of meanings were in a continual state of flux. Hence he argued that there is no set meaning in the text; both the reader and the meanings in the text are changed and transformed through each interaction between reader and text.

These ideas were taken up in a radical way by Derrida (1981) – probably the leading exponent of post-modern thinking. He argued that it is not only language but culturally signifying practices in general that are arbitrary, i.e. they derive their meanings from sets of differences within other discourses. These social practices are known as *texts*. This term is crucial for it refers to more than the traditional notion of a printed book; it can also refer to any form of representing social meanings such as paintings, landscapes, buildings, maps and advertising. The meaning of any text cannot therefore be considered in isola-tion but should be considered in the context of a network of differences. It follows that there can be no definitive underlying structure. A given text is not a static thing but a continually evolving set of meanings. This implies that the consumers of texts are as important as their initial producers. Readers are not therefore simply passive receptors of universal truths passed down from on high by some creative genius. Derrida argues that we do not have stable fixed identities, instead, they are multiple, unstable and continually evolving through every encounter with different discourses – what is termed *intertextuality*.

Whereas modernism has assumed a close relationship between the signifier and the signified, post-modernists reject this, arguing that the two are continually breaking apart. This has also become known as *post-structuralism*. Whereas structuralism assumed some underlying relationship between the sign and its underlying determinant, under post-structuralism they are completely disconnected. This led in literary studies to the method of *deconstruction* – breaking down the intersecting discourses within particular texts.

Many other figures have contributed to the development of cultural studies, but no description of the movement would be complete without mention of two scholars who have been especially important in a British context. The first is Raymond Williams (1971), who did much to broaden our view of culture away from 'high art' to consider the wide variety of ways in which cultural meanings are socially constructed. The second key author is Stuart Hall who, building upon Williams' approach, emphasised the ways in which culture is bound up with power and domination (Hall, 1981). Hence, the language symbols and ritual of everyday life help to reinforce power relations. However, culture is seldom, if ever, homogenous and stable – it is always contested and in flux. This somewhat fragile and contested nature of culture is well expressed by Jenks:

All cultural phenomena, though often formidable in the constraint they exercise, are nevertheless fragile in that they are generated and maintained by virtue of acting members of a society placing and sustaining their own values within them.

(Jenks, 1993, p. 53)

Hall draws attention to the two main traditions of cultural studies. First there is what he terms the *culturalist* paradigm which places emphasis upon the making of culture. Second, there is the *structuralist* approach which emphasises the determining conditions underlying cultural practices (Hall, 1981). However, under Hall's leadership at the Centre for Contemporary Cultural Studies, potential conflicts between these approaches have been ameliorated by a neo-Gramscian concept of hegemony (Jenks, 1993).

CULTURAL STUDIES AND GEOGRAPHY

The division in cultural studies noted by Hall is to some extent mirrored in geography although there is much overlap between approaches. On the one hand, there is emerging a body of work that is centrally concerned with issues of writing and representation (e.g. Barnes and Duncan, 1992). This approach accepts the post-modern notion that our writings, and texts in general, are not some mirror that simply reflects the world around us (the *mimetic* approach). Instead it is argued that our portrayal of the world reflects discourses or sets of meanings that draw upon other texts. Within this perspective the notion of objectivity is simply one type of metaphor or way of looking at the world. This approach tends to take on a post-modern hue to the extent that some writers in the genre reject the notion of some privileged, superior way of looking at the world. Such an approach sees various centres of power as manifest in all types of writing and cultural representation and is unwilling to privilege any one particular perspective. However, relatively few geographers have accepted the full implications of this approach which imply that all forms of knowledge and rationality are relative (Duncan and Ley, 1993).

A second, and hitherto more common, approach has also been concerned with ideologies and discourses but is more concerned to relate these to power relations and material circumstances. This approach may in turn be discerned at two scales. The first is at what one can term the macro-scale and is manifest in attempts to relate cultural change to broader changes in the nature of capitalism. The most famous of these attempts is the work of Jameson (1991). He argues that post-modernism is the cultural logic of late capitalism. Thus, a populace fragmented by a diversity of values, and lacking in the collective institutions to organise around universal themes, fits in well with the needs of global capitalism for acquiescent, passive consumers. Furthermore, he argues that culture itself has become commodified. Thus, images and styles are not some accessory to material goods, they are products in themselves. A similar view is taken by Harvey (1989). He notes the links between an emerging regime of accumulation, which he terms flexible accumulation, and post-modern culture. However, Harvey is cautious about directions of causality or the extent of change; in comparison with Jameson he envisages many more elements of continuity. Nevertheless, Harvey does present an overarching explanation for change and, unlike post-modernists, privileges his Marxian approach.

Another writer dealing with this broad scale is Baudrillard (1988). He argues that post-modern culture is based on images or copies of the real world (known as *simulacra*) which are difficult if not impossible to distinguish from reality. Advertising and mass media produce a set of signs that have their own internal meanings and not those related to external objects, producing instead what he termed a *hyperreality*. Although he set out to provide an explanation linked in a Marxian way to the economic base, Baudrillard rejected this approach, arguing that in the current era it is impossible to develop general theories.

These broad-scale studies of culture have been enormously influential but none of them has much to say directly about the changing character of the welfare state. However, it is not difficult to draw out such inferences. For example, the new language of 'efficiency', 'independence', 'self-reliance' and 'customer' satisfaction represents a whole new set of discourses that are involved in the transformation of the welfare state. At the same time there are large pockets of resistance to these changes in welfare. However, cultural studies has had far more influence upon a second micro-level looking at cultural diversity and institutional change. These studies have derived much of their influence from the work of Foucault who, like Baudrillard, was firmly opposed to all-embracing theories such as Marxism.

Michel Foucault is possibly *the* cult intellectual of the late twentieth century. Like Derrida and Baudrillard, Foucault rejected what he termed totalising theories such as Marxism or Freudian psychoanalysis. He claimed that such theories were always coercive in their practical implications – tending towards forcing people to do things against their will. However, unlike the pessimistic Baudrillard, he did believe that resistance to dominant views was possible. Hence, at numerous periods of his life he was actively involved in a variety of political movements and especially those concerned with prison reform (Macey, 1993).

Foucault's views were derived in large measure from his analysis of some of the key institutions of the welfare state and, in particular, prisons and asylums (Foucault, 1967; 1979). He noted that in earlier traditional societies punishment and control was typically undertaken in public through public executions and beatings. In modern societies, however, punishment and control is more likely to take place behind closed doors in prisons. He extended this idea to conceive of the 'disciplinary society' in which control is exercised through a variety of institutions including not only prisons, but hospitals, schools and factories. He argued that a central issue in these places was the relationship between knowledge and power. For Foucault, power comes from knowledge. Hence, within these institutions there were methods of controlling the inmates that were bolstered by systems of knowledge or discourses, i.e. sets of meanings. A crucial point in Foucault's analysis is that power is not simply imposed in a repressive way from above by the state but it also comes 'from below' and 'from within', through families, social groups and various types of institution. The discourses operating within these settings serve to shape the views that people take of themselves. Power is therefore everywhere, operating through people rather than being imposed upon people from some higher agency. For this reason, space as well as time plays an important part in his analysis. Foucault used the metaphor of the Panopticon – a model plan of a prison devised by Jeremy Bentham in which a tower enables continual surveillance of all prisoners – to describe the surveillance techniques present in the disciplinary society. From a Foucaultian perspective, therefore, power is exercised through everyday encounters and social practices called 'micropowers' (a theme that we noted earlier in Giddens' writings – see Chapter 5). Furthermore, Foucault argued that these discourses took on different meanings in different contexts. This means that these systems of knowledge could not be reduced to broad, class-based forms of exploitation. Nevertheless, the diversity of these discourses means that there are always elements of opposition to the prevailing power relations within places. According to Foucault, power is not simply a negative, external force; it is an essential ingredient in the construction of everyday life. Foucault's approach stresses the pluralism of knowledge but has been criticised for being divorced from some broader, overarching theory of capitalist development (Driver, 1985).

Foucault's work has been extraordinarily influential in a wide range of social sciences. For example, within the field of industrial relations, Knights and Sturdy (1989) used Foucault's work to interpret new forms of computer technology in the finance industry. They argue that new forms of computer technology enable close scrutiny of the performance of individuals, thus ensuring self-discipline on the part of the workers. Workers thus become fragmented, individualised and often demoralised by the intensification of work. Whilst there are crude parallels with the notion of a disciplinary society, Austrin (1994) argues that this is a simplified view of power. Instead, he focuses upon the individual appraisal systems developed within the finance industry. These require individuals to give particular accounts of themselves which are cross checked with other sources of information such as that derived from customer quality surveys. In effect, workers have to engage in a particular discourse or else be disciplined for poor performance. However, he documents the emergence of resistance to change through the actions of trade unions championing the case of particular individuals with grievances. Austrin argues that change in the finance industry has involved a new set of discourses in which the subjects can construct themselves in different ways.

CULTURAL STUDIES AND THE
WELFARE STATE

As stressed in the introduction to this chapter, the 'cultural turn' has had a huge impact upon human geography. Foucault's work in particular has been especially influential (Driver, 1985; Philo, 1989). For example, recognition of the plurality of knowledge has led to talk of geographies rather than *the* geography of a subject. However, there have been relatively few direct applications of Foucault's ideas, as in the industrial relations example cited above; rather, Foucault's ideas have influenced geographers' wider notions about knowledge, power and representation. Furthermore, relatively little of this work has dealt directly with the changing structure of the welfare state.

One notable exception is the work of Painter (1992). He utilises the same material on the diffusion of CCT that was discussed in Chapter 4. However, cast in the wake of cultural studies, the insights are more convincing than those taken from regulation theory. Painter notes six implications that may be derived from the 'cultural turn':

1. There can be no single 'corporate culture' in any organisation. Organisational cultures are inevitably diverse and contested.
2. Cultural change is always political, as well as technical and organisational.
3. Cultures are rarely stable but are constantly subject to renegotiation and transformation.
4. Cultural change is often resisted by groups with differing sets of values.
5. Cultures are not equal but are riven by power and domination.
6. Wide variations in local cultures mean that cultural change cannot always be introduced everywhere in the same manner.

The last point is especially important for geographers and was also noted in Chapter 5. Painter therefore examines some of the *material practices* (organisational structures and the provision of services) and the *discursive practices* (the words, symbols and ideas used to represent the changes) involved in contracting-out. For example, in the London Borough of Wandsworth, a council that was keen to go ahead with contracting-out, local trade unions were portrayed as acting against the interests of consumers. As in the case of contracting-out in the NHS discussed in Chapter 5, the trade unions learned from this portrayal. Subsequently, they encouraged a new discourse that portrayed themselves as protecting the quality of services on behalf of the users of services. The language used here was important; trade unions deliberately referred to 'users of services' to resist all the implications of change that are associated with the word 'customers'.

In contrast, the language used in Manchester City Council revealed an attitude to contracting-out that was very different to Wandsworth. In all literature relating to CCT the council referred to 'enforced tendering' to portray both the staff and the users of services as victims together of an externally imposed policy. Sets of meanings were also generated by the language used by trade unionists involved in setting up the catering contracts for schools. The catering committee, composed of dinner ladies, pupils, parents, teachers and shop stewards, generated a set of meanings that portrayed the various demands of these groups as compatible. The crucial point to emerge in all these examples is that the words are not free floating – they embody cultural values full of power and domination.

Another noteworthy study of institutional change is the work of Halford and Savage (1995). They argue that institutional change is not imposed from above by some rational economic logic. Instead:

restructuring should be understood in terms of the social and cultural practices internal to organisations which construct particular qualities as desirable or undesirable and, therefore, that restructuring is tied up with re-defining and contesting the sorts of personal qualities organisational members are expected to possess.

(Halford and Savage, 1995, p. 98)

They focus in particular upon the gender implications of local authority restructuring. They note that local authorities have traditionally been divided into functionally-based departments dominated by particular professions. This approach was characterised by a strong belief in technical solutions to problems. In addition, there was a culture that tended to stress the authority of the producer of the service over the user. Promotion tended to be given to those who had put in long years of service. In such an environment, women, through their discontinuous career histories (typically taking time out from work to look after children when young), tended to be disadvantaged in terms of promotion. As we have seen in previous chapters, a whole host of changes in recent years have attempted to make local authorities less bureaucratic and paternalistic. Recent managerial changes have placed much greater emphasis upon flexibility and innovation rather than technical competence. It was clear from Halford and Savage's research that these changes were beginning to undermine the male professional ethos that had previously dominated the local authorities they studied. In theory, this should mean greater opportunities for promotion of women and there was some evidence of this. However, many of these newly created posts taken up by women were not part of the main departmental management hierarchy which was still dominated by men. Furthermore, they note evidence of a new type of organisational masculinity based on long hours. However, the outcome of recent developments is uncertain. As with all cultural change, there are diverse values which are the subject of intense struggle and conflict. The Halford and Savage paper therefore raises the issue of gender which has been somewhat neglected in both regulation theory and structuration theory. Studies such as this illustrate the influence of cultural studies in social science, since this has led to a consideration of viewpoints that have tended to be hidden in the past. For example, in recent years feminist perspectives have radically affected our view of the welfare state.

FEMINIST CRITIQUES OF WELFARE STATES

The reactions of feminists to the post-modern turn discussed above have been many and varied. Some have welcomed the increased recognition of difference and diversity whilst others suspect a masculinist plot to subvert feminist critiques, just at the time when they have begun to make a profound impact upon many aspects of social research. Nevertheless, there has been a generally healthy interchange of ideas between post-modernism and feminism (e.g. Deutsche, 1991; Massey, 1991; Bondi and Domosh, 1992). Certainly, feminist critiques have played an increasing role in analysing welfare states (e.g. Borchorst and Siim, 1987; Dale and Foster, 1986; Lewis, 1992; Pateman, 1988; Rose, 1986; Siim, 1990; Ungerson, 1990). There are many examples of the ways in which the structure of welfare states involves both explicit and implicit assumptions about the roles of men and women. The classic example in the British case is the fact that the original proposals for the welfare state inspired by Beveridge assumed that a woman's primary role was as wife and mother. Women were therefore classified as dependent upon their husbands and originally entitled to lower rates of national insurance contributions and benefits. Similarly, until the mid-1980s, the invalid care allowance was not granted to married women caring for a sick or disabled relative on the grounds that this was a part of their normal duties (Williams, 1994). As McDowell argues, the Fordist welfare state was based on the assumption of male labour in the formal workplace under-pinned by unpaid female domestic labour in the home within nuclear families (McDowell, 1991). Although the welfare state is usually regarded as a gain for the most disadvantaged, it is now widely recognised that it was, and is, based in certain respects upon the subordination of women.

Feminists have also directed much critical comment against the regulationist-inspired interpretations of welfare structures. Their basic charge is that this political economy approach is based around notions of class exploitation and thereby either ignores or downplays other social relations based around gender, race, disability, age and sexuality, which all had a crucial impact upon the consti-tution of the welfare state (Jenson, 1990; Williams, 1989; 1993; 1994). It is argued that the notion of the Fordist/post-Fordist welfare state is centred around a white, male, able-bodied experience which tends to marginalise other types of worker who were central to the Fordist welfare state. In particular, female and black migrant workers were a core element underpinning the Fordist welfare state.

Taken together, feminist critiques and the cultural studies movement have highlighted the diversity of groups in society and the ways in which their needs have often been neglected by welfare regimes.

RACISM AND THE WELFARE STATE

Williams (1989) notes that there is not a coherent anti-racist critique of the welfare state comparable with feminist work. It is certainly true that in the US an extensive range of studies have been undertaken to examine if racial minori-ties are systematically discriminated against. However, this work has produced contradictory results and suffers from numerous methodological problems (see Pinch 1985 for a review). In some instances, additional resources have been allocated to meet the additional needs of minority groups but it seems that often these are not sufficient to make up for the disadvantages experienced by racial minorities. Furthermore, there is substantial evidence of systematic discrimination against racial minorities in the operations of welfare states.

It has also been argued that racial minorities, by taking up poorly paid jobs that whites were unwilling to do, were a crucial element in the old welfare state. Bakshi and associates note that the labour movement developed a negative attitude towards immigration and was at the forefront in arguing for immigration controls (Bakshi et al., 1995). Indeed, issues of welfare are still linked with immigration. The 1971 Immigration Act indicates that the wives and children of Commonwealth citizens can enter the country only if a sponsor can support them without recourse to public funds (Bakshi et al., 1995). Recent moves towards competitive tendering have also impacted upon racial minorities since they comprise a high proportion of the workforce in ancillary occupations in some parts of the UK, especially in the major conurbations.

DISABILITY AND WELFARE

In the last decade the disabled have constituted an important type of so-called 'new social movement'. Thus, various disabled groups have become increasingly vocal in petitioning for their special needs (Butler, 1994; Hahn, 1989). In part, this reflects a shift in thinking about the nature of disability. In the past, disability was usually thought of as a physical defect affecting a particular individual. However, in recent years this view has become replaced by a *social constructionist* approach. This latter approach argues that, rather like gender, race and sexuality, disability is a social construction. Hence, disability is connected with the attitudes and structures of oppression imposed by an able-bodied society rather than the failings of particular individuals. It is therefore argued that if sufficient facilities were provided then the notion of disability would largely disappear. Thus, using a wheelchair would be seen as no more of an impairment than the wearing of glasses. There has recently been a growing awareness of the stereo-typical ways in which disabled people are portrayed in the media as victims, unsympathetic, pitiable or simply a problem (see Table 6.1).

French (1993) has argued that this social constructionist argument goes too far, in that it neglects certain intrinsic experiences (such as pain or impaired vision) that are part of the bodily experiences of the disabled. Nevertheless, there has in recent years been the beginning of a sea-change in attitudes towards disability with increased facilities being provided in city centres and new buildings. However, Williams (1994) argues that the recent restructuring of the welfare state has been to the overall detriment of the disabled. General reductions in welfare spending have left many disabled groups neglected, despite all the rhetoric about increasing the diversity of welfare services. For example, concentrating benefits amongst the most needy has led to a focus upon those

Table 6.1 Ten media stereotypes of disabled people

Pitiable and Pathetic: Charity adverts; Children in need; Tiny Tim

Object of Violence: Films like *Whatever Happened to Baby Jane?*

Sinister or Evil: Dr No; Dr Strangelove; Richard III.

Atmosphere: curios in comics, books or films (e.g. *The Hunchback of Notre Dame*).

Triumph over Tragedy: e.g. the last item in the news.

Laughable: the butt of jokes, e.g. Mr Magoo.

Bearing a Grudge: Laura in *The Glass Menagerie*.

Burden or Outcast; the Morlocks in *The X-Men* or in *The Mask*.

Non-sexual or incapable of full relationships: Clifford Chatterley in *Lady Chatterley's Lover*.

Incapable of fully participating in everyday life: absence from everyday situations and not shown as integral and productive members of society.

Source: adapted from *Guardian* 13th October 1995

with long-term disability rather than those disabled by old age. Furthermore, the increased discretion given to social workers to assess the needs of the disabled are seen as a denial of fundamental human rights (Williams, 1994).

SEXUALITY AND WELFARE

One final issue, which has so far not received the attention it deserves is the relationship between sexuality and welfare. Welfare systems tend to be based on assumptions of heterosexual relations within households, with the result that the needs and aspirations of gays and lesbians have tended to be neglected.

On the need side of the equation, families and local communities sometimes ostracise their members who 'come out', putting gays and lesbians under enormous pressure to find independent accommodation. Whereas discrimination and harassment on the grounds of race or gender can be challenged on legal grounds, there is no equivalent legislation on grounds of sexuality (Egerton, 1990). Furthermore, those gays and lesbians who chose to conceal their sexual identity were often put under severe psychological pressure with associated social problems. It has been argued that whereas gay men may gain prestige and material advantage from their independence, single women are often seen in a negative light (Eichenbaum and Orbach, 1982).

On the supply side, there can also be considerable problems. Housing suppliers in both the private and public sectors have tended not to recognise the needs of same-sex households. Given the relatively low wages paid to women, this can often disadvantage lesbians in their efforts to obtain housing in the owner occupied or privately rented sector. However, gay men may also face considerable problems in obtaining mortgages through concerns about HIV and Aids. Widespread homophobia also means that gays and lesbians fear how they will be treated in old age in residential and nursing homes. These complex and important issues surrounding sexuality and welfare demand much greater attention than they have so far received (see Knopp, 1994).

CONCLUSION

At first glance, the cultural studies movement in its most post-modern forms threatens to overthrow the whole 'applecart' of analysis of the welfare state, bringing into question as it does the rationality of the various discourses used to understand the world. However, it could be argued that this influence has been beneficial to the extent that it as it has brought into question the power relations behind many commentaries on welfare. It has also drawn attention to the plurality of viewpoints in society. Thus, a key insight to emerge from the cultural studies movement in recent years is that the meanings of words depend upon the context in which they are being used. The crucial point is that these words become part of a discourse – a wider set of meanings. These meanings have important implications for power and the material relations of life or how people gain access to resources. It is for these reasons that the torrent of new words plays such a crucial role in attempts to restructure the welfare state. Furthermore, the nature of the discourses and power shifts varies considerably between different places. Consequently, the cultural studies movement is likely to play a crucial role in future studies of welfare restructuring.

Yet one might question whether this concern with diversity has not gone too far. For example, how does the desire to take account of the diversity of human aspirations square with the assertion, made at the outset of this book, that people are very similar in their basic desires for adequate shelter, fulfilling sources of employment and decent health care? These issues, together with the changing nature of the welfare state and how we can best understand its changing geography, are considered in the last chapter.

FURTHER READING

There has been a torrent of work on cultural studies flooding through geography and the social sciences in recent years. Unfortunately, some of this work is pretentious and inaccessible (to say the least). Nevertheless, a useful and clear introduction may be found in Jenks (1993). Hall, Held and McGrew (1992) also provide an extremely clear set of readings. Painter (1995) is a clear and concise introduction to these ideas in the context of political geography. As in the case of regulation theory, it is always worth seeking out the original French material (e.g. Foucault, 1967; 1979; Gordon, 1980) although, once again, it is refreshing to find some illuminating commentaries on a complex set of ideas (e.g. Driver, 1985). Harvey (1989) is a lucid introduction to post-modernism and associated issues of 'flexible accumulation'. It is also worth seeking out the introductions in Barnes and Duncan (1992) and Duncan and Ley (1993). Useful perspectives on the relationships between gender and welfare are Lewis (1993), Williams (1989) and McDowell (1991).

7

CONCLUSIONS

This book has examined both the changing nature of welfare states and the changing geographical perspectives upon these changes. It should have become clear by now that these two elements are closely related; not only have welfare states become more diverse, fragmented and pluralistic, but our ways of analysing these changes have also become increasingly diverse. No longer is there a search for one all-embracing way of analysing society, but there is a recognition of the diversity and plurality of value systems and approaches. Similarly, our confidence in producing a master blueprint to solve welfare problems also seems to have evaporated. This final chapter examines the implications of this situation for the geographical analysis of the welfare state.

THE POST-MODERN DILEMMA

Inevitably, things are more complex than the generalisations made in the introduction to this chapter might imply. Despite the diversity of approaches, there are some underlying similarities in recent attempts by geographers to analyse change. Indeed, one might even argue that there are closer bonds between many contemporary geographical studies than existed in the 1970s when there was the division between quantitative and structuralist approaches. For example, there is now a fairly widespread recognition of the need to take account of the role of social struggles and human agency in shaping social outcomes but in the context of particular economic, social and political constraints. Thus, one may dispute whether there is a specific regime of accumulation called post-Fordism, but few would dispute that globalisation has radically altered the context in which particular nation-states have to operate. Similarly, one may dispute the meanings of the language, but few would deny that a whole battery of terms are being used to change the ways in which the institutions of the welfare state operate. There is also a widespread recognition of the ways in which particular geographical contexts, whether they be nation states, regions, cities or particular micro-institutional contexts, can influence human behaviour. Consequently, although some of the details of the terminology of the particular approaches discussed in this book may disappear in the future, they have underlying elements that will endure. One can see this in the continuing influence of structuration theory. Many would not use the Giddensian terminology but many studies are operating within frameworks that have been greatly influenced by his approach. Similar sentiments apply to regulation theory and cultural studies.

Of course, one should not disguise the profound differences that persist between contemporary approaches. Probably the most profound difference at present is between post-modernists who insist on the plurality of knowledge and forms of rationality and those who would attempt to relate changes to some overarching theory of economic and social change. Indeed, from a post-modern perspective, both the positivist and structuralist approaches of the past were 'in the same camp' to the extent that they were both striving for a privileged and superior way of interpreting the world (Barnes and Duncan, 1992). However, as indicated in Chapter 6, there is no clear-cut division between post-modern and modernist approaches, and few geographers have yet fully embraced the full implications of post-modernist thinking. Indeed, most geographers concerned with issues of social welfare seem to be bent upon providing insights that may contribute towards social improvement, albeit in a way that increasingly recognises the diversity of peoples' wishes. It is in the latter spirit that this last chapter is written, for, as Susan Smith has argued, 'Without a better sense of the value of prescription, geography is powerless to challenge the subtle ideologies that legitimise enduring social inequalities'(Smith, 1993, p. 72).

SOCIAL NEEDS AND UNIVERSAL NOTIONS OF JUSTICE

A key issue raised by post-modernist thinking is whether, in an age in which we recognise the plurality of aspirations and value systems, we can have a universally applicable conception of welfare and human need. Certainly, there have been many attempts to define the basic minimum levels of food, clothing and shelter needed to sustain human existence (see Smith, 1994 for a review). A generally held view is that while it is possible to define very basic levels of nutrition and warmth needed to sustain people on the edge of starvation and destitution, in the more advanced economies needs must be relative to the levels of living of the society as a whole. But if needs are diverse and reflect many values or discourses, then must we abandon any attempt at universal notions of justice and provision as originally incorporated into the welfare state? Interestingly, in recent years there have been a number of attempts to challenge the notion that needs are all relative.

One of the most important of these is the work of Doyal and Gough (1991). They argue that most of those who reject absolute notions of need do, in fact, implicitly incorporate such absolute notions in their analysis. For example, those on the New Right who advocate individual self-reliance still hold on to the notion of a safety net for the worse off. Doyal and Gough argue that all people share the desire to avoid physical harm and that this desire translates into two basic needs. The first need is for physical health to continue and to live and function effectively and the second need is for personal autonomy, the ability to make personal choices about how to live their lives. Doyal and Gough then translate these basic goals into intermediate needs of food, water, housing, health care and so on. They argue that while the ways in which we meet these needs are culturally variable, the needs themselves are universal. They then go on to derive measures of these indices of needs for different countries. From this perspective they argue that welfare needs can be measured over time and place and that this approach provides an escape from the impasse of post-modern deconstruction of multiple discourses. Their approach suggests that people will attempt to meet their basic needs in a multiplicity of ways (through consumption of housing, food and health care) but this should not prevent analysis of indices of these basic needs in the first instance such as homelessness, infant mortality rates, life expectancy and so on.

Further support for universal notions of justice come from David Harvey (1992). He notes that the fashion to celebrate difference and multiple discourses tends to undermine the notion of universal conceptions of social justice. He also notes that in any struggle over resources there are usually multiple viewpoints and discourses at work. However, he rejects the view that these are all equally valid. Justice and rationality, he acknowledges, take on different meanings across time and space, but it is important to look at the material basis for producing these differing concepts. He argues that it is the ideas and discourses of the powerful that have the most influence. Whilst acknowledging the need to recognise difference, Harvey locates the current penchant for diversity (what he calls 'fetishising situatedness') in the emerging right-wing agenda of market solutions to solve the problems of the capitalist nations. As we have seen in this book, this involves the notion that the market is the best solution to the problems of both efficiency and distribution. In fact, as Harvey notes, the idea that the market will solve problems is just another way of applying a universal concept of rationality and justice.

Drawing upon the work of Doyal, Gough and Harvey, we can therefore see a substantial research agenda for future work on the welfare state. In particular, this work will monitor whether the market and pseudo-market mechanisms being introduced deliver all or indeed any of the advantages that they promise. Support for such an agenda also comes – somewhat paradoxically, since he writes from a liberal perspective – from Bennett (1989). Bennett has no time for conceptual frameworks that are not part of a validated and tested scheme and he therefore rejects Marxist approaches in geography. However, like Harvey, he rejects the extreme relativism of post-modernism. He argues that no theory can any longer claim privileged status but must instead be supported by its ability to be confirmed by practices and outcomes. Bennett is concerned with three basic questions: first, 'did the policy reach the target or people it was intended to reach in the intended form?'; second, 'did the policy have the intended effects on the behaviour or conditions of the target audience or people it was intended to reach?'; and third, 'did the policy improve society as a whole?' One does not have to share Bennett's particular liberal perspective view to see that these are crucial questions that need addressing in the context of the changing welfare state. Amongst the many questions that require research the following would also seem to be especially important:

1. Which groups (in which areas) are most disadvantaged by the reductions in the scale and scope of the welfare state?
2. Which social groups have suffered most from the emphasis upon self-reliance?
3. To what extent have recent developments either improved or worsened the lot of women, men, ethnic and racial minorities, the disabled, the elderly, gays, lesbians and persons in particular regions?
4. What have been the long-term consequences of asset sales such as local authority housing?
5. Has the policy of privatisation increased consumer choice?

6. Have the early failures of the policy of deinstitutionalisation been remedied or made worse?
7. Have their been any real benefits from investment and technical change in welfare services?
8. To what extent has contracting-out increased the efficiency and effectiveness of service provision?
9. Has functional flexibility increased the quality of public services or has this been a cost-cutting exercise?
10. Have internal markets produced the gains in efficiency and effectiveness that were anticipated?

As usual, geographers will approach these issues sensitive to ways in which social processes operate in different places. Towards this end the various research themes outlined in the second part of this book are useful for various purposes. Thus, regulation theory, despite its flaws, sensitises us to the broader economic context of change and the diversity of post-(or neo-) Fordist processes. Structuration theory, in contrast, directs our attention to the operation of people in both local and national contexts. It highlights the ways in which people can draw upon structures to affect in various ways the global process sweeping throughout welfare states. And finally, cultural studies directs our attention towards the complexity of language and the diversity of meanings that individuals attach to these words. It highlights the multiplicity of forms of rationality and the changing power relations within the changing welfare state.

Although it is too soon to make clear judgements about many of the changes to welfare systems, such as the introduction of internal markets, we do have over fifteen years of experience of neo-liberal reforms and these do enable us to make some final speculations about the shift towards market mechanisms. These issues are considered in the last part of this book.

TOWARDS THE 'MEAN AND LEAN' WELFARE STATE?

It seems that hardly a day passes without some politician or pundit drawing attention to the 'crisis' of the welfare state. Although this crisis has been discussed ever since the western economies encountered severe economic problems in the 1970s, it seems that the strength of the debate and the pace of change has gathered momentum in recent years. For example, there have been numerous assertions in recent years that in the wake of rising expectations, demographic trends and the increasing costs of medical care, the NHS will no longer be able to cope with the demands placed upon it. Another theme has been the huge future costs of long-term care for the elderly. Various radical solutions have been mooted for the NHS including charging for non-core treatments, rationing non-urgent treatments and encouraging patients to adopt private insurance, or at least to pay for extra services. The UK is, of course, not alone in this respect; in the Netherlands, Canada and New Zealand efforts have been made in recent years to curb health sending by focusing upon core treatments. Even in European states in which the welfare state is extensive and broadly supported by a wide section of the population, including the middle classes, such as France and Sweden, politicians have begun to consider ways of reducing the burden of welfare in order to reduce public spending deficits and thereby please international investors. In both the US and the UK it has also become fashionable for those on the moralising end of the New Right to berate single-parent mothers for the various social problems of crime and delinquency.

At the same time, there are numerous examples of needs being neglected. For example, although local authority rents in the UK have increased by over 100 per cent in real terms since 1979, the extra resources produced have not been ploughed back into new public housing The result is a huge backlog of housing in need of repair or replacement. Similarly, a report by the UK Mental Health Foundation in 1994 showed that although over £2000 million had been saved by the closure of large psychiatric hospitals in recent years, this has not been allocated back into community-based facilities. It has been estimated that at least £540 million is needed to resource community care properly. Other studies highlight the patchy nature of services for disturbed young people as local authorities have withdrawn some of their professionals from family mental health teams in the wake of pressures for expenditure cuts. The results has been increased waiting times for first appointments and increased difficulties in obtaining places in therapeutic communities. Of course, needs will always be infinite in scope, and expectations are rising, but there is a huge body of evidence that basic needs are not being met and many more examples could be cited. Many would argue that economic competition from Asian and American industries threatens many of the inefficient and protected European industries, such that unemployment and social needs could escalate considerably in future. However, such has been the pace of change that even left-of-centre political parties have begun to engage in the language of 'workfare', 'partnership', 'community' and 'self-reliance'. Thus, the UK Labour party is reluctant to engage in heavy welfare spending commitments for fear of upsetting both voters and the financiers in the City of London.

In this context we may be able to draw some lessons from the perspectives considered in this volume. Regulation theory, for example, posits that there should be some solution or 'fix' between the needs of the economy and that of consumption. Whatever the failings of the classic Fordist welfare state, there does seem to have been such a fix for a period of time in the 1950s and 1960s. From this perspective it can also be suggested that current arrangements look highly flawed and unlikely to produce long-term stability. For example, the neo-liberal approach adopted in the UK, involving the deregulation of labour markets and the erosion of collective welfare, has created many problems. To begin with, welfare spending has had to increase to cope with paying unemployment benefit. The welfare state evolved after the Second World War in a time when long-term unemployment was relatively uncommon, men were paid a 'working wage' and the majority of women stayed at home; clearly these conditions are no longer as prevalent as they were in the past. At the same time the insecurity of employment has begun to affect all sections of society with the result that they have become reluctant to engage in consumption – the engine needed to keep the capitalist system moving forward. Although some sections of society have been doing well through tax cuts and investments, there is a large and growing section of the population that have become marginalised. The result has been a huge increase in social problems resulting from a lack of social cohesion such as crime, homelessness, ill health, disability, social disorganisation and poor education levels. Furthermore, the restriction of spending on education and training in the UK is not providing the skilled workforce needed for the twenty-first century. Thus, what Jessop (1994) calls the 'neoliberal shell' for post-Fordism is looking incapable of providing many of the factors needed to produce competitive economy in an era of global competition, as is made strikingly clear by transport policy in the UK in recent years.

The obvious remedy for these problems lies in additional taxation and collective provision, yet this a strategy that political parties of both the Left and Right throughout the world have been reluctant to embrace in recent years for fear of devastating political consequences. Of course, whatever may be their responses to questionnaires, the public have in recent years displayed hostility in the ballot box to the idea of paying additional taxes. However, it may not be too optimistic and far-fetched to perceive that we may be at the beginning of a sea-change in attitudes. In recent years the notion that individuals should be entirely responsible for meeting their own welfare needs has been brought into question for many individuals by their experience of the reformed welfare state. For example, meeting the long-term care of the elderly through private funds threatens to wipe out very quickly savings that have been accrued over a lifetime. As is widely known, private insurance schemes are not suited to long-term care of this type. Thus, many have come to the conclusion that such needs can be met only by some form of state organised collective insurance. Arguably, this need not lead to a reversion back to the monolithic welfare systems of the past, since a diversity of approaches are needed to deal with a diversity of needs. Any welfare system now needs to take account of the three Es – equity, efficiency and effectiveness. It is important to realise in this context that the UK currently has one of the lowest proportions of GDP devoted to welfare spending amongst all the nations of western Europe. Furthermore, there are many ways in which states can adjust to the exigencies of global capitalism. For example, it has been argued that within Europe there is an emerging consensus about a new form of corporatist welfare state that is rather different from the British neo-liberal variety (Benington and Taylor, 1993). As discussed in Chapter 1, the welfare states of continental Europe have traditionally had labour market solutions to social problems but these are threatened by rising proportions of the retired and unemployed. However, the reversion to non-state forms of welfare pluralism in Germany and Sweden is taking on a more positive role than in the UK. Rather than being seen as a means of rolling back the state, welfare pluralism has had a more positive role in attempting to ensure greater social inclusion and empowerment for citizens (Beinington and Taylor, 1992; Eyles, 1989). Australia is another nation which in recent years has made a consistent attempt to attack many of the long-established inequalities associated with the traditional welfare state. For example, there has been a consistent attempt to reduce income differentials between men and women (Smith, 1989).

In this context there is a critical task for geographers in helping to create a new form of social 'fix' or 'settlement'. This means that it is not sufficient to simply describe and monitor changes in welfare regimes (however complex a task this may be); it is also important to utilise some of the concepts outlined in this book to create a sustained and coherent critique of recent developments. In conjunction with our 'geographical imaginations' we may then begin to under-stand the diverse worlds of welfare.

FURTHER READING

The contrasting views on welfare of Doyal and Gough (1991), Harvey (1992) and Bennett (1989) are all worth deliberating on in the context of this chapter. In addition, the work of Young (1990) is quite crucial in any consideration of the future of welfare. Smith (1994) provides an excellent overview of the concept of social justice (see also Hay, 1995). Hutton (1995) provides a thought-provoking commentary on many of the issues considered in this chapter. Hutton's book has already proved to be a 'classic' and may have an important impact upon future developments. It is also worth examining Mulgan (1994) and the readings in Oakley and Williams (1994).

GLOSSARY

Definitions can often seen like legal documents, correct in details but incomprehensible to those who are not experts. In contrast, the definitions in the following glossary are designed to be as accessible as possible and to convey the basic elements of the ideas. To understand more about the ideas and concepts listed below, reference should be made to the relevant chapters in this book and the further reading suggested; words in italic indicate that there is a separate entry for that term. If you are familiar with the ideas then such definitions can seem obvious. However, if you are new to a field, such glossaries can provide vital signposts through a complex terrain.

Note: All words in italics are defined below.

Asset sales The sale of publicly owned organisations (such as utilities) and assets (such as public housing) to the private sector (Ascher, 1987). See *privatisation*. Chapter 2.

Civil society All the elements of society outside of government including private sector businesses, the family and the voluntary sector.

Classical Marxism The ideas formulated by Marx in the nineteenth century. Contrast with *neo-Marxism*.

Collective consumption Usually refers to goods and services provided by the public sector. Less often refers to services that literally have to be consumed by a group of people in a collective manner (such as a lecture). The term originated in a Marxian theory formulated by Castells which argues that there are certain services that are crucial for the maintenance of capitalism but that are too expensive for provision by individual capitalist enterprises and therefore require provision through non-market means via the public sector. The theory also attempts to define cities as essentially places for the consumption of public services – a notion that has been much criticised (Pinch, 1985). See *neo-Marxism*, *public goods*.

Commercialisation The tendency for publicly owned organisations to behave like private sector companies (such as through the imposition of user charges). Also termed *proprietarisation*. See also *corporatisation*. Chapter 3.

Commodification The use of private markets rather than public sector allocation mechanisms to allocate goods and services. Also termed *recommodification* and *marketisation*. Chapter 1.

Community care Care for the needy in local communities either in small decentralised facilities or in private households – both supported by teams of community-based professionals. Associated with *deinstitutionalisation*. A policy much criticised for inadequate funding and resources – hence the term care 'in' the community but not 'by' the community. Chapter 2.

Competitive tendering A process through which contracts are awarded on the basis of competitive (usually secret) bidding by a variety of agencies according to specified criteria such as cost, quality and flexibility. See *compulsory competitive tendering*, and *contracting-in*. Chapter 3.

Compulsory competitive tendering (CCT) The imposition by the UK central government of competitive tendering upon local governments and health authorities that were reluctant to adopt such a policy. Chapter 3.

Contextual theory A broad trend in social analysis characterised by a desire to understand the settings or contexts within which human behaviour takes place. These approaches seek to understand how people are influenced by, but at the same create, these contexts. See *situatedness*. Chapter 5.

Contracting-in A situation in which a contract is won by a sub-division of the parent organisation putting the work up for tender. This is said to be kept *in-house*. See also *market testing*. Chapters 3 and 5.

Contracting-out A situation in which one organisation contracts with another external organisation for the provision of a good or service (Ascher, 1987). Often

associated with *competitive tendering* but this need not be the case. May also be termed *sub-contracting*, *distancing* or *outsourcing*. See also *market testing*. Chapters 3 and 5.

Contractualisation The use of contracts to govern the relationships between organisations and sub-divisions within organisations. Increasingly used to allocate public services by allocating contracts to either private sector companies, charitable or voluntary organisations or internal departments within the public sector. See *contracting-in*, *contracting-out*, *internal markets*. Chapters 2, 3 and 5.

Corporate philanthropy Donations and gifts made to the voluntary sector by corporations, often through the creation of separately administered trusts. Chapter 2.

Corporatisation An extreme form of *commercialisation* in which publicly owned organisations behave in an identical manner to private sector companies. Chapter 3.

Corporatism Forms of social organisation in which certain interest groups, usually certain sectors of business and organised labour, have privileged access to government. Characterised by collaboration to achieve economic objectives. See *neo-corporatism*, *industrial achievement model*, *integrated welfare state* and *welfare corporatism*.

Cultural studies A complex set of developments in social analysis which pay attention to the complexity of cultural values and meanings. See *culture* and *'cultural turn'*. Chapter 6.

'Cultural turn' The tendency for many social sciences to pay greater attention to issues of *culture*. Also termed the *linguistic turn* because of the attention given to language and the ways in which ideas are represented. See *post-structuralism*, *deconstruction*. Chapter 6.

Culture This may be broadly interpreted as 'ways of life'. It consists of the values that people hold, the rules and norms they obey and the material objects they use (Giddens, 1989). Also commonly regarded as sytems of shared meanings (see *discourse*). Chapter 6.

Decentralisation The fragmentation and geographical dispersal of service provider units so that they are more accessible to the public. May be associated with *devolution* but the two policies are distinct. See *tapering*. Chapter 3.

Deconstruction A form of analysis that examines the various discourses represented by various forms of representation (known as texts). These meanings are regarded as continually changing through the interactions of the reader/viewer and the text in question. See *discourse*, *text*. Chapter 6.

Deinstitutionalisation The closure of institutions providing long-term care for needy groups and their replacement by various alternative forms of care including purpose-built or converted smaller facilities and care within private households by families supported by teams of community-based professionals such as nurses, doctors and social workers (Dear and Wolch, 1987). See *community care*, *rationalisation*, *reinstitutionalisation*, *self-provisioning* and *domestication*. Chapter 2.

Deregulation Policies designed to increase competition by breaking up state monopolies and introducing a number of private agencies to provide goods and services. May also be applied to the deregulation of labour markets through policies to erode workers' rights and to increase labour flexibility. See *commodification* and *marketisation*. Chapter 3.

Devolution The sub-division of welfare organisations into separate units each with their own budgets. Usually associated with devolution of responsibilities and with enhanced performance monitoring of the units. See also *decentralisation*. Chapter 3.

Differentiated welfare state A welfare state in which social policy is distinct from, and unrelated to, economic and industrial policy. Also referred to as the *pluralist welfare state*. Chapter 1.

Disciplinary society A society in which control is exercised through socialisation processes as manifest in schools, hospitals and factories. See *micropowers*. Chapter 6.

Discourse Sets of meanings that are indicated by various *texts*. See *deconstruction*. Chapter 6.

Discursive practices The words, signs, symbols and ideas that are used to represent *material practices*. Chapter 6.

Distanciation The tendency for interactions and communications between people to be stretched across time and space through the use of books, newspapers, telephones, faxes and the like. Also termed *space-time distanciation*. Chapter 5.

Distancing Another term for *contracting out* – a situation when one organisation contracts with another external organisation for the provision of a good or service. Chapter 3.

Domestication The use of family and household labour (Urry, 1987). Has been forced upon some households (and usually women within them) through the run-down of state provision. See *community care*. Chapter 3.

Double hermeneutic The need for researchers to be aware of their own values as well as those of the people they are studying. See *hermeneutics, situatedness.* Chapter 6.

Economic determinism Theories that attempt to relate social changes directly to underlying economic changes in society and that play down the ability of people to make decisions to affect their destinies. Contrast with *voluntarism.* Chapters 4, 5 and 6.

Economies of scale Factors that cause the average cost of a commodity to fall as the scale of output increases. There are two main types, see *external economies of scale, internal economies of scale.* Chapter 4.

Embeddedness In Chapter 5 of this book this term refers to Pettigrew and associates' (1992) desire to study the interconnections between various levels of analysis. This term is more generally used in social science to assert that economic behaviour is not affected by universal values that are invariant (as in *neo-classical economics*) but is intimately related to cultural values that may be highly specific in time and space. Also termed social embededness. See *culture, situatedness.*

Essentialism The notion that there are basic unvarying elements that determine or strongly affect the behaviour of people and social systems. For example, the idea that there are inherent differences in the behaviour of men and women, or basic immutable laws of economics that govern capitalist societies. Essentialism is the opposite of a *social constructionism.*

Ethnic group A minority group that shares a distinctive *culture.* This is conceptually distinct from the notion of a *racial group* but in practice the two are intimately linked.

Ethnography The study of *culture,* especially the values and norms of minority ethnic groups. Often linked to qualitative research methods such as participant observation and semi- or unstructured questionnaires.

Extensive regime of accumulation A phase of capitalist evolution during which profits were enhanced primarily through increasing the amount of output and expanding the scale of the market rather than through increasing the productivity of workers. A key concept in certain forms of *regulation theory.* Contrast with the *intensive regime of accumulation* and the *flexible regime of accumulation.* Chapter 4.

External economies of scale Factors that reduce the costs of production when the industry to which the firm belongs is large (e.g. the development of specialist suppliers, services and skilled workers). These factors apply irrespective of the size of the individual firm. Contrast with *internal economies of scale.* See also *new industrial spaces, vertical disintegration.* Chapter 4.

Feminism A broad social movement advocating equal rights for men and women. Also various forms of academic analysis that attempt to expose the different attitudes that lead women to be oppressed. See *gender, sexism.* Chapter 6.

Flexibilisation A set of policies designed to increase the capacity of firms to adjust to variations in market demand. May be applied to forms of industrial organisation and to labour practices as well as to both private and public sector bodies (Pinch, 1989). See also *functional flexibility* and *numerical flexibility.* Chapter 3.

Flexible regime of accumulation The idea that the *intensive regime of accumulation* has been replaced by a new regime in which the prime emphasis is upon flexibility of production. See also *regulation theory*, *post-Fordism*, *flexibilisation* and *flexible specialisation*. Chapter 4.

Flexible specialisation The idea that mass production using unskilled workers is being replaced by batch production of specialised products in small companies using skilled workers (Piore and Sabel, 1984). Has similarities with the concept of *post-Fordism* in *regulation theory* but is highly voluntarist in approach and is less concerned with matching industrial change to wider economic forces. See *voluntarism*. Chapter 4.

Fordism A system of industrial organisation established by Henry Ford in Detroit at the beginning of the twentieth century for the mass production of automobiles. In *regulation theory* the concept refers to a *regime of accumulation* which was dominant after the Second World War based on *Keynesianism*, mass production and the welfare state. Chapter 4.

Functional flexibility The capacity of firms (and public sector organisations) to deploy the skills of their employees to match the changing tasks required by variations in workload (IMS, 1986). Chapter 3.

Functionalism A type of reasoning incorporated, either explicitly or implicitly, into a great deal of social theory that is characterised by a number of mistakes. These mistakes include: attributing 'needs' to social systems; assuming that social systems are functionally ordered and cohesive; assuming *teleology* in social systems (i.e. that events can be explained only by movement towards some pre-ordained end); assuming effects as causes; and assuming empirically unverified or unverifiable statements as tautological statements (i.e. true by definition) (Thrift, 1983). May also be used to refer to a form of managerial philosophy

that advocates the sub-division of organisations around particular tasks and responsibilities. Chapters 4, 5 and 6.

Gender Social, psychological and cultural differences between men and women (rather than biological differences of sex). See *feminism, sexism*.

Geographical imagination The need for geographers to understand the diversity of cultural values of those they study in different contexts (and to recognise the influence of their own values upon the frameworks they use to represent these people). See *contextual theory, situatedness*.

Globalisation The tendency for economies and national political systems to become integrated at a global scale. Also the tendency for the emergence of a global *culture*, i.e. universal trends that it is argued are sweeping through all nations. Chapter 1.

Hermeneutics Theories that examine the complexity of people's views, ideas and subjective interpretations of the world around them.

Human agency The capacity of people to make choices and take actions to affect their destinies. Often played down in structuralist and deterministic theories. Contrast with *economic determinism*. See *reflexivity*.

'Hollowing-out' The transfer of powers from the nation state to political units at other levels such as the supra-national or sub-national level (see Jessop, 1994). May also refer to the transfer of powers at the local government level to private sector organisations rather than other political jurisdictions (Patterson and Pinch, P., 1995). Also used to refer to the *contracting-out* of activities by private corporations. Chapter 3.

Hyperreality Sets of signs within forms of representation such as advertising which have internal meanings with each other rather than the some underlying reality. May also be thought of as copies that become more important or take on separate meanings from the originals they represent. See *simulacra*. Chapter 6.

Ideal type A notion derived from ideal type analysis which attempts to simplify and exaggerate key elements of reality for the sake of conceptual and analytic clarity (e.g. as in classification schemes of welfare states). Chapter 1.

Ideological superstructure Sets of institutions such as schools and the family that reinforce ideas that serve the interests of the wealthy and powerful. These are distinguished from the underlying economic base. Also termed *state apparatuses*. See *relative autonomy*. Chapter 1.

Ideology Ideas that support the interests of the wealthy and powerful.

Industrial achievement model A welfare state in which social policy is geared towards the smooth and efficient functioning of the economy (Titmuss, 1974). Similar to the *integrated welfare state*. Chapter 1.

Instanciation The idea that the social structure does not exist 'out there' independently of people but is continually created by people through their everyday interactions. See *structuration theory* and *recurrent social practices*. Chapter 5.

Integrated welfare state A welfare state in which economic and social objectives are integrated (Titmuss, 1974). Similar to *industrial achievement model*. Chapter 1.

Intensification Increases in labour productivity through managerial and organisational changes (Massey, 1984). Chapter 3.

Intensive regime of accumulation. A period of history during which profits were enhanced through increasing the efficiency with which inputs to the production system were used. Also termed *Fordism*. See *regulation theory*, *regime of accumulation*. Chapter 4.

Internal economies of scale Factors that lower the cost of production for a firm irrespective of the size of the industry to which the firm belongs. These factors usually involve large rates of output which lead to the possibility of specialist machines to increase rates of productivity and which thereby help to recoup the costs of installing such machinery. Contrast with *external economies of scale*. See also *Fordism*. Chapter 4.

Internal markets Attempts to introduce market mechanisms within public sector organisations by dividing them up into separate units for the purchase and supply of services and by establishing various contracts and trading agreements between these agencies (Walsh, 1995). Chapter 3.

Intertexuality The continually changing meanings that result from the interactions between the reader/observer and the *text*. Part of a form of analysis known as *deconstruction*. Contrast with *mimetic approach*. Chapter 6.

Investment and technical change Capital investment in new forms of machinery and equipment. Often associated with employment loss (Massey, 1984). Chapter 3.

Joint supply The idea that some goods and services have characteristics such that if they can be supplied to one person, they can be supplied to all other persons at no extra cost. See *theory of public goods*. Chapter 1.

Jurisdictional partitioning The sub-division of nation states into political and administrative units with responsibility for the allocation of goods and services (Pinch, 1985). Chapter 1.

Keynesianism A set of policies that under-pinned welfare states in the 1950s and 1960s. The objective was to manage economies by countering the lack of demand in recessions through government spending – hence the term 'demand management'. This approach was undermined by inflation and high unemployment in the 1970s. Chapter 1.

Keynesian welfare state (KWS) A welfare state underpinned by Keynesian demand management policies. Also charac-terised by universal benefits, citizens' rights and increasing standards of provision. See also *Keynesianism* and *welfare statism*. Chapters 1 and 4.

Laissez faire An ideology that underpinned many capitalist societies in the nineteenth century which argued that the state should not intervene in the operation of private markets.

Liberalism A set of ideas that underpin the western democracies. Characterised by a belief in the value of the individual whose rights should not be subordinated to those of society as a whole; tolerance for opposing views; and a belief in equality of opportunity rather than equality of outcomes. See also *neo-liberalism*, *libertari-anism* and *New Right theory*.

Libertarianism A form of New Right theory that argues that, apart from preserving property rights, the state should leave individuals to do whatever they wish (Nozick, 1974; Smith 1994). See also *neo-liberalism*. Chapter 1.

Locales Distinctive settings or contexts in which interactions between people take place. See *structuration theory*, *contextual theory* and *recursiveness*. Chapter 5.

Managerialism A type of analysis that focuses upon the influence of managers upon access to resources and services. Also known as urban managerialism. These managers are also known as 'social gatekeepers' and 'street level bureaucrats' (Pinch, 1985).

Marketisation Transferring the allocation of goods and services from non-market to market principles. See *internal market*, *commodification*. Chapter 2.

Market testing A process whereby various external organisations are invited to bid for contracts by an organisation wishing to test the efficiency of its own internal division in supplying the good or service in question. See *contracting-in*, *contracting-out*. Chapter 3.

Material practices Organisational struc-tures that are concerned with the manu-facture, distribution and consumption of goods and services. Contrast with *discursive practices*. Chapter 6.

Merit goods Goods and services that are regarded as so desirable that they cannot be left to private markets and are allocated by the public sector. The reason for this is that the benefits to the community exceed those to the individual, so that the latter will tend to consume too little for the common good. Chapter 1.

Metanarrative A theory or conceptual framework that purports to be a superior way of looking at the world, providing superior or privileged insights. Also known as a *totalising narrative*. See also *post-modernism*, *deconstruction*. Chapter 6.

Micropowers Everyday interactions through which social control becomes exercised. See *recurrent social practices*. Chapter 6.

Mimetic approach The idea that writing and other forms of representation are mirrors that reflect the world around us. Chapter 6.

Mixed economies of welfare A system in which welfare needs are met by a diverse set of agencies, including the voluntary and private sectors, rather than exclusively by the state (Pinker, 1992). Also termed *welfare pluralism*. Chapter -2.

Mode of regulation An idea central to *regulation theory* that asserts that conflicts within a capitalist society are mediated by various types of norms, rules and regulations that are manifest in various types of legislation and institutions. See also *regime of accumulation*. Chapter 4.

Mode of societalisation Various institutional values that help to integrate a *regime of accumulation* with a *mode of regulation* (Jessop, 1994). So far, this notion has not been formulated in depth. Chapter 4.

Monetisation The introduction of a monetary system to facilitate the exchange of goods and services. Chapter 1.

Neo-classical economics Attempts to update the ideas of the classical economists of the late eighteenth and early nineteenth centuries. Characterised by a belief in the value of market mechanisms. The approach tends to focus upon microlevel individual market problems rather than wider economic issues. It looks for universal, unchanging principles of human economic behaviour and tends to ignore the social context of economic activity. Contrast with *embeddedness* and *situatedness*.

Neo-corporatism Corporatist forms of social organisation designed to increase the competitiveness of the economy (Jessop, 1994). See *Shumpeterian workfare state*.

Contrast with *neo-liberalism* and *neo-statism*. Chapter 4.

Neo-Fordism Various strategies designed to overcome the problems inherent in the Fordist regime of accumulation but without fundamentally transforming it (Aglietta, 1979). This may be regarded as a transition period until a new *regime of accumulation* emerges. Chapter 4. See *Fordism, regulation theory, mode of regulation*.

Neo-liberalism Strategies to make economies competitive by various types of *New Right* policy including *privatisation* and *deregulation*. Contrast with *neo-corporatism* and *neo-statism*. (May sometimes be referred to as neo-classical liberalism.) Chapter 4.

Neo-Marxism Attempts to upgrade classical Marxist theories in the light of development in social theory and society in the twentieth century. Also termed Marxian theories. Chapter 4.

Neo-statism Direct state intervention to achieve international competitiveness (Jessop, 1994). Contrast with *neo-corporatism* and *neo-liberalism*. Chapter 4.

New industrial spaces The geographical concentration of new small firms in the same sector involved in dense networks of sub-contracting and collaboration. Also termed 'industrial districts'. Chapter 4.

New Right A set of ideas that share in common a belief in the superiority of market mechanisms as the most efficient means of ensuring the production and distribution of goods and services (Dunleavy, 1991). Chapter 4.

'New Wave' management theory A set of ideas that stress the advantages of demolishing elaborate managerial hierarchies and their replacement by 'leaner, flatter' managerial structures (Osbourne and Gaebler, 1992). Often associated with *devolution*. Chapter 3.

Non-excludability The idea that some goods and services have characteristics such that it is impossible to withhold them from those who do not wish to pay for them. See *theory of public goods, non-rejectability*. Chapter 1.

Non-rejectability The idea that some goods and services have characteristics such that once they are supplied to one person, they must be consumed by all, even those who do not wish to do so. See *theory of public goods*, *non-excludabilty*. Chapter 1.

Not-for-profit sector A term often used in the US to denote the charitable or *voluntary sector*. Chapter 2.

Numerical flexibility The ability of firms (and public sector organisations) to adjust their labour inputs over time to meet variations in output. May be in the form of temporary, part-time or casualised forms of working. Chapter 3.

Pluralist welfare state A welfare system in which social and economic objectives are, or tend to be, unrelated. Also termed the *differentiated welfare state* (Titmuss, 1974). Chapter 1.

Positionality The values adopted by an individual. Linked to 'standpoint theory' which argues that writings are not an objective mirror of reality but reflect the cultural context in which they are produced. Contrast with the *mimetic approach*. See *contextual theory* and *situated knowledge*.

Post-Fordism A new *regime of accumulation* based around flexibility which it is assumed has replaced, or is about to replace, the Fordist *regime of accumulation* based on mass production. Similar to *flexible accumulation*. Contrast with *Fordism* and *neo-Fordism*. Also used more generally to refer to lower level concepts such as labour practices and forms of industrial organisation. Chapter 4.

Post-modernism A broad trend in social thinking that rejects the idea that there is one superior way of understanding the world (see *metalanguage* and *totalising narrative*). Strongly linked to a type of analysis known as *deconstruction*. May also be regarded as a style characterised by eclecticism, irony and pastiche (as in architecture but also in writing and advertising). Other interpretations are as a period of history and a cultural trend, which is the logical accompaniment to the era of *post-Fordism* or *flexible accumulation*. Chapter 6.

Post-structuralism A type of analysis. Unlike *structuralism*, which assumes a close relationship between the *signifier* and the *signified*, post-structuralism assumes that these are disconnected and in a continual state of flux. See *deconstruction*, *text* and *intertextuality*. Chapter 6.

Privatisation A diverse set of policies designed to introduce private ownership and/or private market allocation mechanisms to goods and services previously allocated and owned by the public sector. See *asset sales, commercialisation, commodification* and *marketisation*. Chapter 2.

Proprietarisation The tendency for voluntary or non-profit agencies to adopt the strategies of private sector organisations (McLafferty, 1989). See *commercialisation*. Chapter 2.

Public goods Goods and services that have characteristics that make it impossible for them to be allocated by private markets (Samuelson, 1954; Musgrave, 1958). May also be used in a general sense to indicate goods and services provided by the public sector. See *theory of public goods*. Chapter 1.

Racial group A group of people who are assumed to be biologically distinct because of some characteristic of physical appearance, usually skin colour. Since these differences are of no greater significance than other physical attributes such as hair colour, a racial group is one in which certain physical attributes are selected as being ethnically significant (Giddens, 1989). See *racism, ethnic group*.

Racism A set of ideas and social practices that ascribe negative characteristics to a particular *racial group* who are mistakenly assumed to be biologically distinct. See *ethnic group*.

Rationalisation A term much used in economic geography to refer to the closure of industrial capacity which has also been used to refer to the closure of facilities within the welfare state (Massey, 1984; Pinch, 1989). Chapter 2.

Recomodification The reallocation of goods and services from non-market to market mechanisms. Similar to *marketisation, commodification*.

Recursiveness A key element of *structuration theory* which recognises that social systems are made up of the numerous everyday interactions of people. Also termed *recurrent social practices*. Chapter 5.

Redistributive welfare state A welfare system in which services are allocated on some universal criteria of social need (Titmuss, 1974). Chapter 1.

Reflexivity The capacity of people to have knowledge of the situations that face them and to make choices based on this knowledge. See *human agency*. Chapter 5.

Regime of accumulation An abstract concept central to *regulation theory* which argues that from time to time within capitalist societies there emerge stable sets of arrangements that serve to link production and consumption. See also *mode of regulation, Fordism, neo-Fordism* and *post-Fordism*. Chapter 4.

Regulation theory A set of Marxist-inspired ideas that attempt to relate changes in labour practices and forms of industrial and social organisation to wider economic developments and the changing relations between nation-states. Chapter 4.

Reinstitutionalisation The process whereby ex-patients of closed institutions such as psychiatric hospitals end up in other forms of institution, especially prisons (Dear and Wolch, 1987). Chapter 3.

Relative autonomy The idea embodied in certain structuralist approaches that the ideological superstucture is not rigidly determined by the economic base of society. See also *economic determinism, functionalism, economic superstructure, state apparatuses*. Chapter 6.

Reserve army of labour The idea that within capitalist economies there are pools of workers who are given employment in times of high demand and laid off in times of recession (Braverman, 1974). Chapter 4.

Residual welfare state A welfare system that comes into operation only as a last resort when other means of meeting welfare needs, through families, voluntary bodies and private sector agencies, fail (Titmuss, 1974). Chapter 1.

Residualisation Reductions in welfare spending so that services are limited to deprived minorities. Chapter 2.

Self-provisioning A situation where individuals make their own arrangement to meet their welfare needs rather than relying upon the state. The alternatives could be self-help, the voluntary sector or private sector agencies. Derived from the study of changing services (Gershuny and Miles, 1983). See also *domestication*. Chapter 2.

Semiology The study of signs and their meanings. Also termed *semiotics*. See *signifiers*, the *signified* and *text*. Chapter 6.

Sexism Sets of ideas, attitudes and behaviour that ascribe one of the sexes with inferior characteristics. See *gender*, *feminism*.

Shadow state The tendency for the *voluntary sector* to take over services that were previously allocated by the state. The shadow state is diverse and outside of traditional channels of democratic control (Wolch, 1989). Chapter 2.

Shumpeterian workfare state (SWS). An emerging form of welfare state in which the needs of individuals are subordinated to enhancing the international competitiveness of the economy (Jessop, 1994). Unlike the *Keynesian welfare state* the SWS tends to be based on discretion, minimalism and means testing. Chapter 4.

Signified The cultural meaning that is indicated by the *signifier*. See also *text*. Chapter 6.

Signifier That which points to some wider cultural meaning. This can be a word, sign or material object. See the *signified*. Chapter 6.

Simulacra Images or copies of the 'real' world that are difficult to distinguish from the original reality they purport to represent (Baudrillard, 1988). May be thought of as copies that take on a 'life of their own'. A key element in post-modern culture. See *post-modernism*, *hyperreality*. Chapter 6.

Situatedness An approach that recognises that all writings and other forms of representation emerge from people with particular values and in cultures which are distinct in time and space. An approach that denies that there are invariant patterns of human behaviour across time and space, as assumed in much *neo-classical economics*. Also referred to as situated knowledge.

Social constructionism A theory that asserts that most of the differences between people are the result not of their inherent characteristics but of the way in which they are treated by others in society (e.g. because of inadequate facilities, physical incapacity in a wheelchair is regarded as a disability but short-sightedness is not). Can be applied to differences related to ethnicity and gender. See *racism* and *sexism*. Contrast with *essentialism*. Chapter 6.

Social division of labour The social characteristics of the people who undertake different types of work (e.g. age, ethnicity, gender). See also *technical division of labour*. Chapter 4.

Social engineering The belief that society can be improved by rational comprehensive planning based on scientific principles (as in comprehensive slum clearance and urban redevelopment schemes). Chapter 1.

Social reproduction All the various elements that are necessary to reproduce the workforce and the consumers needed to keep a capitalist society functioning (e.g. the family, schools, health services, welfare state, etc.). A key part of Marxian theories which stress the role of the welfare state in overcoming the problems of capitalism (for a review see Pinch, 1985). Much criticised in the past for *functionalism*.

Social movements Pressure groups and organisations with varying degrees of public support petitioning for change, often outside of conventional political channels. These formed an important part of Castell's theory of *collective consumption*.

Space–time distanciation The tendency for interactions and communications between people to be stretched across time and space through the use of newspapers, telephones, faxes and the like. Also termed *distanciation*. See *structuration* theory. Chapter 5.

State apparatuses A term used within structuralist theories to refer to key elements of the ideological superstructure such as the church, family and education system. See *structuralism*. Chapter 6.

Structuralism A theoretical approach derived originally from the study of languages that involves delving below the surface appearance of human activity to examine the underlying structures that affect human behaviour. Chapters 4, 5 and 6.

Structuration theory A theory expounded by Giddens that attempts to bridge the divide between voluntarist and determinist theories. See *voluntarism* and *economic determinism*. Chapter 5.

Structure A key part of *structuration theory* which refers to the rules, norms and resources that individuals draw upon to carry out their lives. See *system, recursiveness*. Chapter 5.

Sub-contracting A situation in which one organisation contracts with another for the provision of a good or service. Also termed *contracting-out*. Chapter 3.

System A term used in many different ways according to the approach in question but is generally used to refer to the interdependent parts of a larger entity. In structuration theory the system is the outcome of all the actions undertaken by people. See *structuration theory, structure, recursiveness, reflexivity*. Chapter 5.

Tapering The tendency for those who live furthest away from the sources of goods and services to consume them less often. This is usually attributed to the increased travel costs or the increased time involved in visiting the source of supply. Also known as the distance-decay effect. Chapter 1.

Taylorism A set of ideas, developed by US engineer Frederick Taylor to manage the labour process, that were adopted by Henry Ford to mass produce automobiles. Also termed the 'principles of scientific management'. These involved simplification of tasks, control of workers and the utilisation of 'time and motion' studies to determine the most efficient ways of working. See *Fordism*. Chapter 4.

Technical division of labour The types of work that need to be undertaken within an industrial system. Contrast with *social division of labour*. Chapter 4.

Text A key concept in *cultural studies* that refers to any form that represents social meanings – not just the written word but paintings, landscapes and buildings. See *discourse* and *deconstruction*. Chapter 6.

Thatcherism A set of policies associated with the Conservative Administrations led by Margaret Thatcher in the UK between 1979 and 1992. The nature of these policies, their cohesiveness and the extent to which they reflected New Right ideas has been the source of much controversy (Jessop et al., 1988). Chapters 1 and 4.

Theory of public goods A theory that states that some goods and services have characteristics that make it impossible for them to be allocated by private markets (Samuelson, 1954; Musgrave, 1958). See *joint supply*, *non-rejectability* and *non-excludability*. Chapter 1.

Totalising narrative A theory that purports to be a privileged way of interpreting the world providing superior insights. May also be termed a *metanarrative*. Chapter 6.

Vertical disintegration A situation in which companies and organisations subcontract work out to other (usually small) organisations. Contrast with *vertical integration*. Chapter 4.

Vertical integration A structure in which functions are integrated into a large organisation in a complex, interdependent hierarchy. Contrast with *vertical disintegration*. Chapter 4.

Voluntarism The use of the voluntary sector to meet welfare needs. See also *shadow state*. This term may also refer to a type of social analysis that envisages people as capable of making decisions to evolve in an almost infinite range of possible directions. This approach therefore plays down the constraints upon people. See also *human agency*. Contrast with *economic determinism*. Chapters 2, 4, 5 and 6.

Voluntary sector A diverse set of non-profit making agencies attempting to meet welfare needs. This sector includes charities, charitable trusts and pressure groups. Chapters 1 and 2.

Welfare corporatism A society charac-terised by corporatist forms of collaboration in which certain groups can gain privileged access to government to derive benefits of various types (e.g. contracts, tax concessions). Usually applies to big business or organised labour rather than the most deprived. See *corporatism*.

Welfare pluralism A system in which welfare needs are met by a diverse set of agencies, including those from the voluntary and private sectors, rather than relying upon universal provision by state agencies (Johnson, 1987). This has a different meaning to the *pluralist welfare state*. Chapter 1.

Welfare regimes The theory that welfare states can be grouped according to the extent to which they possess three sets of characteristics: liberalism, conservatism and social democracy (Esping-Anderson, 1990). Chapter 1.

Welfare state A set of institutions and social arrangements designed to assist people when they are in need because of factors such as illness, unemployment and dependency, through youth and old age. Chapter 1.

Welfare statism The notion that the state should have the responsibility to ensure an adequate standard of living for its citizens through policies designed to achieve full employment, relatively high minimum wages, safe working conditions and income transfers from relatively affluent majorities to deprived minorities (Pfaller et al., 1991). Chapter 1.

Workfare A system in which those who are unemployed have to undertake work in order to receive benefits. Chapter 4.

BIBLIOGRAPHY

Aglietta, M. (1979) *A Theory of Capitalist Regulation*, London: New Left Books.

Allen, J. (1988) 'Fragmented firms, disorganised labour?', in J. Allen and D. Massey (eds) *The Economy in Question*, London: Open University Press.

Allen, J. (1992) 'Post industrialism and post-Fordism', in S. Hall, D. Held and T McGrew (eds) *Modernity and Its Futures*, Cambridge: Polity Press.

Althusser, L. (1966) *For Marx*, London: Verso Books.

Althusser, L. (1971) *Lenin and Philosophy and Other Essays*, London: New Left Books.

Amin, A. (ed.) (1994) *Post-Fordism: A Reader*, Oxford: Blackwell.

Amin, A. and Robins, K. (1990) 'The re-emergence of regional economies? The mythical geography of flexible accumulation', *Environment and Planning D: Society and Space* 8, 7–34.

Ascher, K. (1987) *The Politics of Privatisation*, Basingstoke: Macmillan.

Austrin, T. (1994) 'Positioning resistance and resisting position: human resource management and the politics of appraisal and grievance hearings', in J. M. Jermier, D. Knights and W. R. Nord (eds) *Resistance and Power in Organisations*, London: Routledge.

Bagguley, P. (1991) 'Post-Fordism and enterprise culture: flexibility, autonomy and changes in economic organisation', in R. Keat and N. Abercrombie (eds) *Enterprise Culture*, London: Routledge.

Bakshi, P., Goodwin, M., Painter, J. and Southern, A. (1995) 'Gender, race and class in the local welfare state: moving beyond regulation theory in analysing the transition from Fordism', *Environment and Planning A* 27, 1555–1576.

Barnes, T. J. and Duncan, J. S. (1992) *Writing Worlds*, London: Routledge.

Barthes, R. (1973) *Mythologies*, St. Albans: Paladin.

Baudrillard, J. (1988) *Selected Writings*, Cambridge: Polity Press.

Benington, J. and Taylor, M. (1993) 'Changes and challenges facing the UK welfare state in the Europe of the 1990s', *Policy and Politics* 21, 121–143.

Bennett, R. J. (1980) *The Geography of Public Finance*, London: Methuen.

Bennett, R. J. (1989) 'Whither models and geography in a post-welfarist world?', in B. Macmillan (ed.) *Remodelling Geography*, Oxford: Blackwell.

Bondi, L. (1987) 'School closures and local politics: the negotiation of primary school rationalisation in Manchester', *Political Geography Quarterly* 6, 203–224.

Bondi, L. (1988) 'Political participation and school closures: an investigation of bias in local authority decision-making', *Policy and Politics* 16, 41–54.

Bondi, L. and Domosh, M. (1992) 'Other figures in the landscape: on feminism, post-modernism and geography', *Environment and Planning D: Society and Space* 10, 199–213.

Borchorst, A. and Siim, B. (1987) 'Women and the advanced welfare state – a new kind of patriarchal power', in A. S. Sasoon (ed.) *Women and the State*, New York: Century Hutchinson.

Boyne, G. and Powell, M. (1991) 'Territorial justice: a review of theory and evidence', *Political Geography* 10, 262–281.

Bradford, M. (1995) 'Diversification and division in the English education system: towards a post-Fordist model?', *Environment and Planning A* 27, 1595–1612.

Bradshaw, J. (1972) 'The concept of social need', *New Society* 496, 640–643.

Braverman, H. (1974) *Labor and Monopoly Capital*, New York: Monthly Review Press.

Bryant, G. A. and Jary, D. (1991) 'Introduction: coming to terms with Anthony Giddens', in G. A. Bryant and D. Jary (eds) *Giddens' Theory of Structuration: A Critical Appreciation*, London: Routledge.

Burrows, R. and Loader, B. (eds) (1994) *Towards a Post-Fordist Welfare State?*, London: Routledge.

Butler, R. (1994) 'Geography and the vision-impaired and blind populations', *Transactions of the Institute of British Geographers* 19, 366–368.

Charlesworth, J., Clarke, J. and Cochrane, A. (1995) 'Managing local mixed economies of care', *Environment and Planning A* 27, 1419–1436.

Clark, G. L. (1990a) 'Location, corporate strategy and workers' pensions', *Environment and Planning A* 22, 17–37.

Clark, G. L. (1990b) 'Restructuring, workers' pension rights and the law', *Environment and Planning A* 22, 149–168.

Clark, G. L. (1992) 'Real regulation: the administrative state', *Environment and Planning A* 24, 615–627.

Clarke, S. (1988) 'Overaccumulation, class struggle and the regulation approach', *Capital and Class* 38, 59–92.

Cochrane, A. (1991) 'The changing state of local government: restructuring for the 1990s', *Public Administration* 69, 281–302.

Cochrane, A. and Clarke, J. (eds) (1993) *Comparing Welfare States: Britain in International Context*, London: Sage.

Cohen, I. J. (1989) *Structuration Theory: Anthony Giddens and the Constitution of Social Life*, London: Macmillan.

Cohen, P. (1993) 'Managing hotel services: moving the goalposts', *Health Services Journal* Special Report, 13 May, pages 1–3.

Cousins, C. (1987) *Controlling Social Welfare*, Brighton: Wheatsheaf.

Cousins, C. (1988) 'The restructuring of welfare work: the introduction of General Management and Contracting Out of ancillary services in the NHS', *Work, Employment and Society* 2, 210–228.

Craib, I. (1992) *Anthony Giddens*, London: Routledge.

Crook, A. D. H. (1986) 'Privatisation of housing and the impact of the Conservative Government's initiative on low cost home ownership and private renting between 1979 and 1984 in England and Wales 1: the privatisation policies', *Environment and Planning A* 18, 639–659.

Cubbin, J., Domberger, S. and Meadowcroft, D. (1987) 'Competitive tendering and refuse collection: identifying the sources of efficiency', *Fiscal Studies* 8, 49–58.

Curtis, S. (1989) *The Geography of Public Welfare Provision*, London: Routledge.

Dale, J. and Foster, P. (1986) *Feminists and State*

Welfare, London: Routledge and Kegan Paul.

Dear, M. and Moos, A. (1986) 'Structuration theory in urban analysis: 2 empirical application', *Environment and Planning A* 18, 351–373.

Dear, M. J. and Wolch, J. R. (1987) *Landscapes of Despair: From Deinstitutionalisation to Homelessness*, Cambridge: Polity Press.

Derrida, J. (1981) *Writing and Difference*, London: Routledge.

Deutsche, R. (1991) 'Boys town', *Environment and Planning D: Society and Space* 9, 5–30.

Doyal, L. and Gough, I. (1991) *A Theory of Human Need*, London: Macmillan.

Driver, F. (1985) 'Power, space and the body: a critical assessment of Foucault's "Discipline and Punish"', *Environment and Planning D: Society and Space* 3, 425–446.

Duncan, J. S. and Ley, D. (1993) *Place/Culture/Representation*, London: Routledge.

Duncan, S. and Goodwin, M. (1988) *The Local State and Uneven Development*, Cambridge: Polity Press.

Dunford, M. (1990) 'Theories of regulation', *Environment and Planning D: Society and Space* 8, 297–321.

Dunleavy, P. (1991) *Democracy, Bureaucracy and Public Choice*, Polity Press: Cambridge.

Dunn, R., Forrest, R. and Murie, A. (1987) 'The geography of council house sales in England 1979–85', *Urban Studies* 24, 47–59.

Egerton, J. (1990) 'Out but not down: lesbians' experience of housing', *Feminist Review* 36, 78–88.

Eichenbaum, L. and Orbach, S. (eds) (1982) *Outside In and Inside Out*, Harmondsworth: Penguin.

Elam, M. J. (1990) 'Puzzling out the post-Fordist debate: technology, markets and institutions', *Economic and Industrial Democracy* 11, 9–39.

Esping-Anderson, G. (1990) *The Three Worlds of Welfare Capitalism*, Cambridge: Polity Press.

Exworthy, M. (1993) 'A review of recent structural changes to district health authorities as purchasing agents', *Environment and Planning C: Government and Policy* 11, 279–289.

Eyles, J. (1989) ' Privatising the health and welfare state; the western European experience', in J. Scarpaci (ed.) *Health Services Privatisation in Industrial Societies*, New Brunswick: Rutgers University Press.

Finch, J. (1984) Community care: developing non-sexist alternatives', *Critical Social Policy* 4, 6–18.

Finch, J. (1990) 'The politics of community care', in C. Ungerson (ed.) *Gender and Caring: Work and Women in Britain and Scandinavia*, Brighton: Harvester.

Flynn, R. (1995) *Managerialism, professionalism and quasi-markets*, paper presented at conference entitled 'Professional Autonomy and Managerial Challenges in the Public Sector' at the University of Southampton, 26 May.

Forrest, R. and Murie, A. (1986) 'Marginalisation and subsidised individualism: the sale of council housing in the restructuring of the British welfare state', *International Journal of Urban and Regional Research* 10, 46–66.

Forrest, R. and Murie, A. (1991) *Selling the Welfare State: The Privatisation of Public Housing*, London: Routledge.

Foucault, M. (1967) *Madness and Civilisation*, London: Tavistock.

Foucault, M. (1979) *Discipline and Punish*, London: Penguin.

French, S. (1993) 'Disability, impairment or somewhere in-between?', in J. Swain, V. Finkelstein, S. French and M. Oliver (eds) *Disabling Barriers – Enabling Environments*, London: Sage.

Geiger, R. K. and Wolch, J. (1986) 'A shadow state? Voluntarism in metropolitan Los Angeles', *Society and Space* 4, 351–361.

Gershuny, J. I. and Miles, I. D. (1983) *The New Service Economy*, London: Francis Pinter.

Giddens, A. (1979) *Central Problems in Social Theory: Action Structure and Contradiction in Social Analysis*, Basingstoke: Macmillan.

Giddens, A. (1981) *A Contemporary Critique of Historical Materialism Volume 1* Basingstoke: Macmillan.

Giddens, A. (1984) *The Constitution of Society*, Cambridge: Polity Press.

Giddens, A. (1989) *Sociology*, Cambridge: Polity Press.

Giddens, A. (1991) 'Structuration theory: past, present and future', in G. A. Bryant and D. Jary (eds) *Giddens' Theory of Structuration: A Critical Appreciation*, London: Routledge.

Giddens, A. (1995) *A Contemporary Critique of Historical Materialism, Vol. 2: The Nation State and Violence*, Cambridge: Polity Press.

Glennerster, H. (1995) *The Welfare State in Britain Since 1945*, London: Blackwell.

Goodwin, M., Duncan, S. and Halford, S. (1993) 'Regulation theory, the local state, and the transition of urban politics', *Environment and Planning D: Society and Space* 11, 67–88.

Goodwin, N. and Pinch, S. (1995) 'Explaining geographical variations in the contracting-out of NHS ancillary services: a contextual approach', *Environment and Planning A* 27, 1397–1418.

Gordon, C. (ed.) (1980) *Power/Knowledge: Selected Interviews and Other Writings 1972-1977 Michel Foucault*, New York: Pantheon Books.

Gramsci, A. (1973) *Selections From the Prison Notebooks* (translated by O. Hoare and G. Nowell-Smith), London: Lawrence and Wishart.

Gregson, N. (1986) 'On duality and dualism: the case of time geography and structuration', *Progress in Human Geography* 10, 184–205.

Gregson, N. (1987) 'Structuration theory: some thoughts on the possibilities for empirical research', *Environment and Planning D: Society and Space* 5, 73–91.

Hagerstrand, T. (1973) 'The domain of human geography', in R. J. Chorley (ed.) *Directions in Geography*, London: Methuen.

Hahn, H. (1989) 'Disability and the reproduction of bodily images: the dynamics of human appearances', in J. Wolch, and M. Dear (eds) *The Power of Geography*, Boston: Unwin Hyman.

Hakim, C. (1990) 'Core and peripheries in employers' workforce strategies: evidence from the 1987 ELUS survey', *Work, Employment and Society* 4, 157–188.

Halford, S. and Savage, M. (1995) 'Restructuring organisations, changing people: gender and restructuring in banking and local government', *Work, Employment and Society* 9, 97–122.

Hall, S. (1981) 'Cultural studies: two paradigms', in T. Bennett, G. Martin, C. Mercer and J. Woollacott (eds) *Culture Ideology and Social Process*, London: Open University Press.

Hall, S., Held, D. and McGrew, T. (1992) *Modernity and its Futures*, Cambridge: Polity Press.

Hamnett, C. (1989) 'The political geography of housing in contemporary Britain', in J. Mohan (ed.) *The Political Geography of Contemporary Britain*, Basingstoke: Macmillan.

Harrison, S. and Wistow, G. (1992) 'The purchaser/provider split in English health care: towards explicit rationing?', *Policy and Politics* 20, 123–130.

Harvey, D. (1989) *The Condition of Postmodernity*, Oxford: Blackwell.

Harvey, D. (1992) 'Social justice, postmodernism and the City', *International Journal of Urban and Regional Research* 16, 588–601.

Hay, A. M. (1995) 'Concepts of equity, fairness and justice in geographical studies', *Transactions of the Institute of British Geographers* 20, 500–508.

Hayek, von A. F. (1956) *The Road to Serfdom*, Chicago: University of Chicago Press.

Hazell, R. and Whybrew, T. (1993) *Resourcing the Voluntary Sector: The Funders' Perspective*, London: Association of Charities Foundation.

Held, D. and Thompson, J. B. (eds) (1989) *Social Theory of Modern Societies: Anthony Giddens and His Critics*, Cambridge: Cambridge University Press.

Hirst, P. and Zeitlin, J. (1991) 'Flexible specialisation vs. post-Fordism: theory, evidence and policy implications', *Economy and Society* 20, 1–56.

Hoggart, K. (1985) 'Political control and the sale of local authority dwellings, 1974–1983', *Environment and Planning C: Government and Policy* 3, 463–474.

Hoggett, P. (1987) 'A farewell to mass production? Decentralisation as an emergent private and public sector paradigm', in P. Hoggett and R. Hambleton (eds) *Decentralisation and Democracy*, Bristol: School for Advanced Urban Studies.

Hoggett, P. (1994) 'The politics of the modernization of the UK welfare state' in R Burrows and B. Loader (eds) *Towards a Post-Fordist Welfare State?*, London: Routledge.

Hudson, B. (1992) 'Quasi-markets in health and social care in Britain: can the public sector respond?', *Politics and Policy* 20, 131–142.

Hudson, R. (1989) 'Labour market changes and new forms of work in old industrial regions: maybe flexibility for some but not flexible accumulation', *Environment and Planning D: Society and Space* 7, 5–30.

Hurd, H., Mason, C. and Pinch, S. (1995) 'The geography of corporate philanthropy in the UK', mimeo, University of Southampton.

Hutton, W. (1995) *The State We're In*, London: Cape.

Institute of Manpower Studies (1986) *New Forms of Work Organisation*, Falmer: University of Sussex.

Jameson, F. (1984) 'Post-modernism, or the cultural logic of late capitalism', *New Left Review* 146, 53–92

Jameson, F. (1984) *Post-modernism, or the Cultural Logic of late Capitalism*, Durham: Duke University Press.

Jenks, C. (1993) *Culture*, London: Routledge.

Jenson, J. (1990) 'Different but not exceptional: the feminism of permeable Fordism', *New Left Review* 184, 58–68.

Jessop, B. (1989) 'Conservative regimes and the transition to post-Fordism' in M. Gottdiener and N. Kominos (eds) *Capitalist Development and Crisis Theory: Accumulation, Regulation and Restructuring*, London: Macmillan.

Jessop, B. (1990) 'Regulation theories in retrospect and prospect', *Economy and Society* 19, 153–216

Jessop, B. (1991) 'The welfare state in the transition from Fordism to Post-Fordism', in B. Jessop, H. Kastendiek, K. Nielson and O. K. Pedersen (eds) *The Politics of Flexibility*, Aldershot: Edward Elgar.

Jessop, B. (1992) 'Fordism and post-Fordism: a critical reformulation', in A. J. Scott and M. Storper (eds) *Pathways to Industrialization and Regional Development*, London: Routledge.

Jessop, B. (1993) 'Towards a Schumpeterian welfare state? Preliminary remarks on a post-Fordist political economy', *Studies in Political Economy* 40, 7–39.

Jessop, B. (1994) 'The transition to post-Fordism and the Schumpeterian workfare state', in R. Burrows and B. Loader (eds) *Towards a Post-Fordist Welfare State?*, London: Routledge.

Jessop, B. (1995) 'Post-Fordism and the state',

in A. Amin (ed.) *Post-Fordism: A Reader*, Oxford: Blackwell.

Jessop, B., Bonnett, K., Bromley, S. and Ling, T. (1988) *Thatcherism: A Tale of Two Nations*, Cambridge: Polity Press.

Johnson, N. (1987) *The Welfare State in Transition: The Theory and Practice of Welfare Pluralism*, Brighton: Wheatsheaf.

Johnston, R. J. (1993) 'The rise and decline of the corporate-welfare state; a comparative analysis in a global context', in P. J. Taylor (ed.) *Political Geography of the Twentieth Century: A Global Analysis*, London: Belhaven.

Jones, K. and Moon, G. (1987) *Health, Disease and Society*, London: Routledge.

Joseph, A. and Phillips, D. (1984) *Accessibility and Utilisation: Geographical Perspectives on Health Care Delivery*, London: Harper and Row.

Kleinman, M. and Whitehead, C. (1987) 'Local variations in the sale of council houses in England, 1979–84', *Regional Studies* 21, 1–12.

Knights, D. and Sturdy, A. (1989) 'New technology and the self-disciplined worker in insurance', in M. Mitchell, J. Varcoe, and S. Yearly (eds) *Deciphering Science and Technology*, London: Macmillan.

Knopp, L. (1994) 'Social justice, sexuality and the city', *Urban Geography* 15, 644–660.

Langan, M. and Clarke, J. (1994) 'Managing the mixed economy of care', in J. Clarke, A. Cochrane and E. McLaughlin (eds) *Managing Social Policy*, London: Sage.

Laws, G. (1988) 'Privatisation and the local welfare state: the case of Toronto's social services', *Transactions of the Institute of British Geographers* 13, 449–465.

Laws, G. (1989) 'Privatization and dependency on the local welfare state', in J. Wolch and M. Dear (eds) *The Power of Geography*, Boston: Unwin Hyman.

Le Grand, J. (1984) 'The future of the welfare state', *New Society* 7 June, 385–386.

Le Grand, J. (1994) *Quasi-Markets and Social Policy*, Studies in Decentralisation and Quasi-Markets, Working paper No.1. Bristol: School of Advanced Urban Studies.

Le Grand, J. and Bartlett, W. (eds) (1993) *Quasi-Markets and Social Policy*, London: Macmillan.

Le Grand, J. and Robinson, R. (eds) (1984) *Privatisation and the Welfare State*, London: Allen and Unwin.

Le Grand, J. and Robinson, R. (eds) (1994) *Evaluating the NHS Reforms*, London: King's Fund Institute.

Lévi-Strauss, C. (1969) *Totemism*, Harmondsworth: Penguin.

Lewis, J. (1992) 'Gender and the development of welfare regimes', *Journal of European Social Policy* 2, 7–32.

Lewis, J. (ed.) (1993) *Women and Social Policies in Europe: Work, Family and the State*, Aldershot: Edward Elgar.

Lipietz, A. (1988) 'Accumulation, crises and the ways out: some methodological reflections on the concept of "regulation"', *International Journal of Political Economy* 18, 10–43.

Lloyd, T. (1993) *The Charity Business*, London: John Murray.

Lovering, J. (1991) 'Fordism's unknown successor: a comment on Scott's theory of flexible accumulation and the re-emergence of regional economies', *International Journal of Urban and Regional Research* 159–174.

Lowndes, V. and Stoker, G. (1992a) 'An evaluation of neighbourhood decentralisation: Part 1: customer and citizen perspectives', *Policy and Politics* 20, 47–61.

Lowndes, V. and Stoker, G. (1992b) 'An evaluation of neighbourhood decentralisation: Part 2: staff and councillor perspectives', *Policy and Politics* 20, 143–152.

Macey, D. (1993) *The Lives of Michel Foucault*, London: Vintage.

Martin, R. (1990) 'Flexible futures and post-

Fordist places: comments on Pathways to Industrialisation and Regional Development in the 1990s – an international conference', *Environment and Planning A* 22, 1276–1280.

Maruo, N. (1986) 'The development of the welfare mix in Japan' in R. Rose and R. Shiratori (eds) *The Welfare State East and West*, New York: Oxford University Press.

Maslow, A. (1954) *Motivation and Personality*, New York: Harper and Row.

Massey, D. (1984) *Spatial Divisions of Labour*, London: Macmillan.

Massey, D. (1991) 'Flexible sexism', *Environment and Planning D: Society and Space* 9, 31–57.

McDowell, L. (1989) 'Women in Thatcher's Britain', in J. Mohan (ed.) *The Political Geography of Contemporary Britain*, Basingstoke: Macmillan.

McDowell, L. (1991) 'Life without father and Ford: the new gender order of post-Fordism', *Transactions, Institute of British Geographers* 16, 400–416.

McGegor, A. and Sproull, A. (1992) 'Employers and the flexible workforce', *Employment Gazette*, May, 225–234.

McLafferty, S. (1989) 'The politics of privatisation, state and local politics and the restructuring of hospitals in New York City', in J. Scarpaci, (ed.) *Health Services Privatisation in Industrial Societies*, New Brunswick: Rutgers University Press.

Meegan, R. (1988) 'A crisis of mass production?', in J. Allen and D. Massey (eds) *The Economy in Question*, London: The Open University Press.

Mishra, R. (1984) *The Welfare State in Crisis*, Brighton: Wheatsheaf.

Mohan, J. (1989) 'Commercialisation and centralisation: towards a new geography of health care', in J. Mohan (ed.) *The Political Geography of Contemporary Britain*, Basingstoke, Macmillan.

Mohan, J. (1995) *A National Health Service? The Restructuring of Health Care in Britain since 1979*, Basingstoke: Macmillan.

Moon, G. and Parnell, R. (1986) 'Private sector involvement in local authority service industry', *Regional Studies* 20, 253–266.

Moos, A. I. and Dear, M. J. (1986) 'Structuration theory in urban analysis: 1 theoretical exegesis', *Environment and Planning A* 18, 231–252.

Moulaert, F. and Swyngedouw, E. (1989) 'A regulation approach to the geography of flexible production systems', *Environment and Planning D: Society and Space* 7, 327–345.

Moulaert, F., Swyngedouw, E. and Wilson, P. (1988) 'Spatial responses to Fordist and post-Fordist accumulation and regulation', *Papers of the Regional Science Association* 64, 11–23.

Mulgan, G. (1994) *Politics in an Anti-political Age*, Oxford: Blackwell.

Musgrave, R. A. (1958) *The Theory of Public Finance*, New York: McGraw Hill.

Nozick, R. (1974) *Anarchy, State and Utopia*, New York: Basic Books.

Oakley, A. and Williams, A. S. (eds) (1994) *The Politics of the Welfare State*, London: UCL Press.

Offe, C. (1984) *Contradictions of the Welfare State*, London: Hutchinson.

Osbourne, D. and Gaebler, T. (1992) *Reinventing Government: How the Entrepreneurial Spirit is Transforming the Public Sector*, Reading MA: Addison Wesley.

Painter, J. (1990) 'Compulsory competitive tendering in local government: the first round', *Public Administration* 69, 191–220.

Painter, J. (1991a) 'The geography of trade union responses to local government privatisation', *Transactions, Institute of British Geographers* 16, 214–226.

Painter, J. (1991b) 'Regulation theory and local government', *Local Government Studies* November/December, 23–44.

Painter, J. (1992) 'The culture of competition', *Public Policy and Administration* 7, 58–68.

Painter, J. (1995) *Politics, Geography and 'Political Geography'*, London: Arnold.

Pateman, C. (1988) 'The patriarchal welfare state', in A. Gutmann (ed.) *Democracy and the Welfare State*, Princeton: Princeton University Press.

Patterson, A. and Pinch, P. L. (1995) '"Hollowing out" the local state: compulsory competitive tendering and the restructuring of British public sector services', *Environment and Planning A* 27, 1437–1462.

Pettigrew, A., McKee, L. and Ferlie, E. (1988) 'Understanding change in the NHS', *Public Administration* 66, 297–317.

Pettigrew, A., Ferlie, E. and McKee, L. (1990) *Managing Strategic Change in the NHS: Final Report*, Warwick: Centre for Corporate Strategy and Change, Warwick Business School, University of Warwick.

Pettigrew, A., Ferlie, E. and McKee, L. (1992) *Managing Strategic Change*, London: Sage.

Pfaller, A., Gough, I. and Therborn, G. (eds) (1991) *Can the Welfare State Compete?*, London: Macmillan.

Phillips, D. R. and Vincent, J. (1986) 'Private residential accommodation for the elderly: geographical aspects of development in Devon', *Transactions of the Institute of British Geographers* 11, 155–173.

Philips, D. R., Vincent, J. and Blacksell, S. (1987) 'Spatial concentration of residential homes for the elderly: planning responses and dilemmas', *Transactions of the Institute of British Geographers* 12, 73–83.

Philo, C. (1989) '"Enough to drive one mad"; the organisation of space in 19th century lunatic asylums', in J. Wolch and M. Dear (eds) *The Power of Geography*, Boston: Unwin Hyman.

Phipps, A. G. (1993) 'An institutional analysis of school closures in Saskatoon and Windsor', *Environment and Planning A* 25, 1607–1626.

Pierson, C. (1991) *Beyond the Welfare State?*, Cambridge: Polity Press.

Pinch, S. (1985) *Cities and Services: The Geography of Collective Consumption*, London: Routledge.

Pinch, S. (1989) 'The restructuring thesis and the study of public services', *Environment and Planning A* 21, 905–927.

Pinch, S. (1994) 'Labour flexibility and the welfare state: is there a post-Fordist model?, in R. Burrows and B. Loader (eds) *Towards a Post-Fordist Welfare State?*, London: Routledge.

Pinch, S. and Storey, A. (1992) 'Flexibility, gender and part-time work: evidence from a survey of the economically active', *Transactions of the Institute of British Geographers* 17, 188–214.

Pinch, S., Mason, C. and Witt, S. (1991) 'Flexible employment strategies in British industry: evidence from the UK "sunbelt"', *Regional Studies* 25, 207–218.

Pinker, R. (1986) 'Social welfare in Japan and Britain: a comparative view. Formal and informal aspects of care', in E. Oyen (ed.) *Comparing Welfare States and their Futures*, Aldershot: Gower.

Pinker, R. (1992) 'Making sense of the mixed economy of welfare', *Social Policy and Administration* 26, 273–284.

Piore, M. and Sabel, C. (1984) *The Second Industrial Divide*, New York: Basic Books.

Pred, A. (1984) 'Place as historically contingent process: structuration and the time geography of becoming places', *Annals of the Association of American Geographers* 74, 279–297.

Pulkingham, J. (1992) 'Employment restructuring in the health service: efficiency initiatives, working practices and workforce composition', *Work, Employment and Society* 6, 397–421.

Ridley, N. (1988) *The Local Right: Enabling not Providing*, London: Centre for Policy Studies.

Roobeek, A. J. M. (1987) 'The crisis of Fordism and the rise of a new technological paradigm', *Futures* 19, 129–154.

Rose, H. (1986) 'Women and the restructuring of the welfare state', in E. Oyen (ed.) *Comparing Welfare States and Their Futures*, Aldershot: Gower.

Rose, R. (1986) 'Common goals but different roles: the state's contribution to the welfare mix', in R. Rose and R. Shiratori (eds) *The Welfare State East and West*, New York: Oxford University Press.

Rose, R. and Shiratori, R. (1986) 'Welfare in society: three worlds or one?', in R. Rose and R. Shiratori (eds) *The Welfare State: East and West*, New York: Oxford University Press.

Rustin, M. (1994) 'Flexibility in higher education', in R. Burrows and B. Loader (eds) *Towards a Post-Fordist Welfare State?* London: Routledge.

Salter, B. (1993) 'The politics of purchasing in the National Health Service', *Policy and Politics* 21, 171–184.

Samuelson, P. A. (1954) 'The pure theory of public expenditure', *The Review of Economics and Statistics* 36, 387–389.

Sayer, A. (1989) 'Post-Fordism in question', *International Journal of Urban and Regional Research* 13, 666–696.

Scott, A. (1986) 'Industrial organisation and location: division of labour, the firm and spatial process', *Economic Geography* 62, 215–231.

Scott, A. (1988) 'Flexible production systems and regional development; the rise of new industrial spaces in North America and Western Europe', *International Journal of Urban and Regional Research* 12, 171–185.

Scott, A. (1991) 'Flexible production systems: analytical tasks and theoritcal horizons – a reply to Lovering', *International Journal of Urban and Regional Research* 15, 130–134.

Scott, A. and Paul, A. (1990) 'Collective order and economic coordination in industrial agglomerations: the technopoles of southern California', *Environment and Planning C: Government and Policy* 8, 179–193.

Scott, A. and Storper, M. (1992) 'Regional development reconsidered', in H. Ernste and V. Meier (eds) *Regional Development and Contemporary Industrial Response*, London: Belhaven.

Sheftner, M. (1980) 'New York fiscal crisis: the politics of inflation and retrenchment', in C. Levine (ed.) *Managing Fiscal Stress*, Chatham, NJ: Chatham House.

Siim, B. (1990) 'Women and the welfare state; between private and public dependence', in C. Ungerson (ed.) *Gender and Caring: Work and Women in Britain and Scandinavia*, Brighton; Harvester.

Smith, C. J. (1981) 'Urban structure and the development of natural support systems for service dependent populations', *The Professional Geographer* 33, 457–465.

Smith, C. and Giggs, J. (eds) (1987) *Location and Stigma*, London: Allen and Unwin.

Smith, D. (1994) *Geography and Social Justice*, Oxford: Blackwell.

Smith, S. (1989) 'Society, space and citizenship: a human geography for the "new times"?', *Transactions of the Institute of British Geographers* 14, 144–156.

Smith, S. (1993) ' Social landscape: continuity and change', in R. J. Johnston (ed.) *A Changing World*, Oxford: Basil Blackwell.

Stoker, G. (1989) 'Creating a local government for a post-Fordist society: the Thatcherite project?', in Stewart, J. and Stoker, G. (eds) *The Future of Local Government*, Basingstoke: Macmillan.

Storper, M. (1985) 'The spatial and temporal constitution of social action: a critical reading of Giddens', *Environment and Planning D: Society and Space* 3, 407–424.

Storper, M. and Scott, A. J. (1989) 'The geographical foundations and social regulation of flexible production complexes' in J. Wolch and M. Dear (eds) *The Power of Geography*, Boston, MA: Unwin Hyman.

Stubbs, J. G. and Barnett, J. R. (1992) 'The geographically uneven development of privatisation: towards a theoretical approach', *Environment and Planning A* 24, 1117–1135.

Taylor, M. (1988) *Into the 1990s: Voluntary Organisations and the Public Sector*, London: National Council for Voluntary Organisations and the Institute for Public Administration.

Taylor, S. (1989) 'Community exclusion of the mentally-ill', in J. Wolch and M. Dear (eds) *The Power of Geography*, Boston: Unwin Hyman.

Therborn, G. (1987) 'Welfare state and capitalist markets', *Acta Sociologica* 30, 237–54.

Thrift, N. (1983) 'On the determination of social action in space and time', *Environment and Planning D: Society and Space* 1, 23–57.

Thrift, N. (1985) 'Bear and mouse or bear and tree? Anthony Giddens's reconstitution of social theory', *Sociology* 19, 609–623.

Thrift, N. (1993) 'The arts of the living, the beauty of the dead: anxieties of being in the work on Anthony Giddens', *Progress in Human Geography* 17, 111–121.

Tickell, A. and Peck, J. A. (1992) 'Accumulation, regulation and the geographies of post-Fordism: missing links in regulationist research', *Progress in Human Geography* 16, 190–218.

Titmuss, R. M. (1974) *Social Policy*, London: Allen and Unwin.

Ungerson, C. (1990) *Gender and Caring: Work and Women in Britain and Scandinavia*, Brighton: Harvetser.

Urry, J. (1987) 'Some social and spatial aspects of services', *Society and Space* 5, 5–26.

Walby, S. (1989) 'Flexibility and the sexual division of labour', in S. Wood (ed.) *The Transformation of Work?*, London: Unwin Hyman.

Walsh, K. (1995) *Public Services and Market Mechanisms: Competition, Contracting and the New Public Management*, Basingstoke: Macmillan.

Warrington, M. J. (1995) 'Welfare pluralism or shadow state? The provision of social housing in the 1990s', *Environment and Planning A* 27, 1341–1360.

Webster, B. (1985) 'A woman's issue: the impact of local authority cuts', *Local Government Studies* 11, 19–46.

Williams, F. (1989) *Social Policy: A Critical Introduction*, Cambridge: Polity Press.

Williams, F. (1993) 'Gender, race and class in British welfare policy', in A. Cochrane and J. Clarke, (eds) *Comparing Welfare States: Britain in International Context*, London: Sage.

Williams, F. (1994) 'Social relations. welfare and the post-Fordism debate', in R. Burrows and B. Loader (eds) *Towards a Post-Fordist Welfare State?*, London: Routledge.

Williams, K., Cutler, A., Williams, J. and Haslam, C. (1987) 'The end of mass production?', *Economy and Society* 16, 405–439.

Williams, R. (1971) *Culture*, London: Fontana.

Wolch, J. R. (1980) 'The residential location of the service-dependent poor', *Annals of the Association of American Geographers* 70, 330–341.

Wolch, J. R. (1981) 'The location of service-dependent households in urban areas', *Economic Geography* 57, 111–129.

Wolch, J. R. (1989) 'The shadow state: transformations in the voluntary sector', in J. Wolch and M. Dear (eds) *The Power of Geography*, Boston MA: Unwin Hyman.

Wolch, J. R. (1990) *The Shadow State: Government and Voluntary Sector in Transition*, New York: The Foundation Centre.

Wolch, J. R. and Geiger, R. K. (1983) 'The

urban dimension of voluntary resources; an exploratory analysis', *Environment and Planning A* 15, 1067–1082.

Wolch, J. R. and Geiger, R. K. (1986) 'Urban restructuring and the not-for-profit sector', *Economic Geography* 62, 3–18.

Wolch, J. R. and Reiner, T. (1985) 'The not-for-profit sector in stable and growing regions', *Urban Affairs Quarterly* 20, 485–510.

Wood, S. J. (1989) 'Review article: New wave management?', *Work, Employment and Society* 3, 379–402.

Young, I. M. (1990) *Justice and the Politics of Difference*, Princeton: Princeton University Press.

INDEX

Printed in the United States
by Baker & Taylor Publisher Services